The Palgrave Concise Historical Atlas of THE BALKANS

Dennis P. Hupchick *and*
Harold E. Cox

palgrave

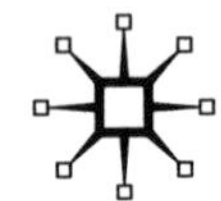

PALGRAVE CONCISE HISTORICAL ATLAS OF THE BALKANS

First published 2001 by **PALGRAVE™**
175 Fifth Avenue, New York, N.Y. 10010 and
Houndmills, Basingstoke, Hampshire RG21 6XS.
Companies and representatives throughout the world.

PALGRAVE™ is the new global publishing imprint of St. Martin's Press LLC Scholarly and Reference Division and Palgrave Publishers Ltd (formerly Macmillan Press Ltd).

ISBN 0-312-23961-0 hardback
ISBN 0-312-23970-X paperback

Library of Congress Cataloging-in-Publication Data
available from the Library of Congress.

A catalogue record of this book is available from the British Library.

Original concept, base maps, and texts by Dennis P. Hupchick.
Finished maps and text graphics by Harold E. Cox.

First edition: September 2001
10 9 8 7 6 5 4 3 2 1

Printed in the United States of America.

Contents

Preface vii
General Key to the Maps viii

INTRODUCTORY MAPS

1. Physical
2. Political, 2001
3. Natural Resources
4. Demographic
5. Cultural

ERA OF BYZANTINE HEGEMONY, 600-1355

6. The East Roman Balkans, Late 6th Century
7. Avar, Slav, and Bulgar Invasions, 7th Century
8. Rise of the First Bulgarian Empire, 7th-10th Centuries
9. Fall of the First Bulgarian Empire, Mid-10th–Early 11th Centuries
10. Rise of Medieval Croatia, 10th-12th Centuries
11. The Balkans, Late 12th Century
12. Crusades in the Balkans, Late 11th-Early 13th Centuries
13. The Balkans after the Fourth Crusade, 1204-1214
14. Byzantium Resurrected, 1261-1328
15. Rise of the Romanian Principalities, Mid-13th–14th Centuries
16. Rise of Medieval Bosnia, 13th-14th Centuries
17. Rise of Medieval Serbia, 13th–Mid-14th Centuries
18. Political Fragmentation, Mid-14th Century

ERA OF OTTOMAN DOMINATION, 1355-1804

19. Ottoman Expansion in the Balkans, Mid-14th–Early 16th Centuries
20. Fall of Constantinople, 1453 (and Ottoman Istanbul)
21. Apex of Ottoman Expansion, Mid-16th Century
22. Ottoman *Millet* Organization, Mid-16th–17th Centuries
23. Habsburg Croatian-Slavonian Military Border, 17th-18th Centuries
24. The Ottoman Balkans, Late 17th-18th Centuries

ERA OF ROMANTIC NATIONALISM, 1804-1878

25. Emergence of Modern Balkan States, 1804-1862
26. The Balkan Crisis of 1875-1876
27. The "San Stefano" Balkans, March 1878
28. The "Berlin" Balkans, July 1878

ERA OF NATION-STATE NATIONALISM, 1878-1944

29. Balkan State Territorial Expansion, 1881-1886
30. The Macedonian Question
31. The Balkans, 1908
32. Bosnia-Hercegovina, 1908-1914
33. The Balkan Wars, 1912-1913
34. World War I in the Balkans
35. The Post-Versailles/Lausanne Balkans, 1923
36. Yugoslavia, 1929-1941
37. Post-Trianon Romania, 1920-1938

38. The Transylvanian Question
39. Bulgaria, 1919-1940
40. Greece, 1923-1941
41. Albania, 1921-1939
42. The Balkans, 1939-1940
43. World War II–The 1941 Balkan Campaign
44. The Axis-Dominated Balkans, 1941-1944

ERA OF COMMUNIST DOMINATION, 1944-1991

45. Balkan Cominform States, 1945-1947
46. The Greek Civil War, 1946-1949
47. Splits in Communism, 1948-1960
48. Collapse of Communism, 1989-1991

ERA OF POST-COMMUNISM, 1991-PRESENT

49. Wars of Yugoslav Succession, 1991-1995
50. The Kosovo Crisis, 1999

Selected Bibliography 115
Index 119

Preface

The tumultuous and oftentimes tragic human events that erupted in the Balkan Peninsula since the collapse of communism in 1989-91 have captured the Western world's attention over the past decade. Unfortunately, many of the West's interventions to halt the violence and uphold basic humanitarian principles in the post-Communist Balkans were hampered by misunderstandings of the conditions on the ground—the background and nature of the issues being fought over by the contending sides. The result has been uneasy cessations of the violence imposed only by Western military presences that bear the prospect of being temporary. Although a number of factors may account for this situation, one undoubtedly has been a limited awareness of Balkan history among Westerners. Fortunately, some scholarly and journalistic books on the subject recently have appeared.

As anyone who attempts to acquire an understanding will discover, the history of the Balkans is lengthy and complex, extending over a millennium and involving the interplay of three civilizations, five empires, three major religions, ten modern nation-states, and some fourteen "major" ethnic groups. Given such historical complexity, we believe that the student and interested general reader can benefit greatly from a set of basic, simple-to-comprehend visual aids—historical maps—depicting important geopolitical stages in the intricate web of Balkan history. For this reason, we offer the following concise historical atlas of the Balkans.

This atlas follows the same format as our successful previous effort, *A Concise Historical Atlas of Eastern Europe* (St. Martin's, 1996; 2nd ed., Palgrave, 2001). Once again, our primary purpose is to provide users with a basic and affordable visual tool for grasping the geopolitical situation at selected important moments in Balkan history. It does not attempt, nor is it intended, to offer a comprehensive overview of every aspect of Balkan history. Thus, there are no specialized maps, tables, or charts treating economic patterns, urbanization, vegetation, land use, annual rainfall, education, transportation, industrial development, demographic movements, and other topics. Nor does it pretend to offer *definitive* cartographic representation of the periods and events covered. It must be understood that a work of this kind sometimes unintentionally can be misleading to the historical neophyte. Geopolitical maps require the presentation of states bounded by borders, but hard and fast state "borders," as we know them today, did not evolve until the late 18th century. Therefore, the user should bear in mind that the state borders appearing in maps depicting situations prior to that time are intended to provide an approximation of the territories controlled by the various state authorities, and that the authorities' territorial control within those states may have ranged from direct to nominal at any given time.

The maps are rendered in two- rather than full-process color. Only those elements deemed necessary for a general understanding of the topics presented are included. Most rivers and mountain ranges, therefore, either do not appear or do so relative to their informational purpose within any given map. Our decisions concerning both scope and presentation were based on considerations of the fundamental purpose of the atlas—basic geopolitical information—and cost.

Each map is accompanied by a page of text. The individual texts are intended to provide a broad perspective on the particular periods and issues represented in the maps. They are not meant to be mere descriptions of specific map elements. Yet, since the concise atlas will be used best as a supplemental resource by the student or general reader, the texts do not constitute a truly comprehensive history of the Balkans when taken as a whole. Numerous factual gaps and lapses exist within and among the texts. Likewise, space limitations do not allow explanations for every foreign or specialized term used in the texts. In both cases, it is assumed that such information will be available to the user from sources outside of this publication.

Regarding spelling, most foreign common terms and proper names appearing in the atlas are rendered in or near their native spellings. Exceptions to this approach are: (1) terms generally known to English speakers in their Anglicized forms (such as the names of states, certain cities, and geographic elements); and (2) the first names of Greek, Russian, and German individuals. Place-names generally are given in their contemporary forms. Although some scholars may take issue with this decision, it should be stated that, in cases other than blatantly ahistorical instances (such as calling Constantinople "Istanbul" or Adrianople "Edirne" prior to their Ottoman conquest), using historical place-names, while technically accurate, has little import for anyone other than specialists. For the student and general reader to whom this atlas is addressed, such name changes tend to be more confusing than informational.

Two approaches are taken to transliterating into Latinized form Slavic terms that natively are written in the Cyrillic alphabet. A "phonetical" system is used for Bulgarian and Russian (for example: **ч** is rendered as **ch**; **ш** as **sh**; **ц** as **ts**; **й** as **i**; **ж** as **zh**; **я** as **ya**, etc.). In the cases of Serbian and Macedonian (Cyrillic languages in pre-1991 Yugoslavia), a "linguistic" system using diacritical marks with some characters is employed (for example: **ч** is rendered as **č**; **ћ** as **ć**; **ш** as **š**; **ц** as **c**; **й** as **j**; **ж** as **ž**; **я** as **ja**, etc.), which is based on the Latin, "Croat" form of Serbo-Croatian commonly used in the West for transliterating "Yugoslav" languages. Turkish terms are spelled in the Latin characters currently used in Turkey with the appropriate diacritical marks.

We wish to thank Michael Flamini, vice president and editorial director at Palgrave, his assistant Amanda Johnson and her editorial staff, and Alan Bradshaw, along with his production staff, for their thoughtful and creative input and overall support of our efforts.

Dennis P. Hupchick
Harold E. Cox
Wilkes University
Wilkes-Barre, PA

General Key to the Maps

Color is employed in this atlas as a tool to make it more "user-friendly" than if it were printed in black and white. On the maps, colored lines and shaded areas highlight important geopolitical developments and help to simplify complex situations that might otherwise be confusing for the user.

International borders	**- - - - - - - - - - - - -**
Regional boundaries	- - - - - - - - - - - - -
Names of states	**YUGOSLAVIA**
Names of regions	*MACEDONIA*
Names of ethnic groups	*Ruthenians*
Names of rivers	*Danube R.*
Names of cities	Vienna

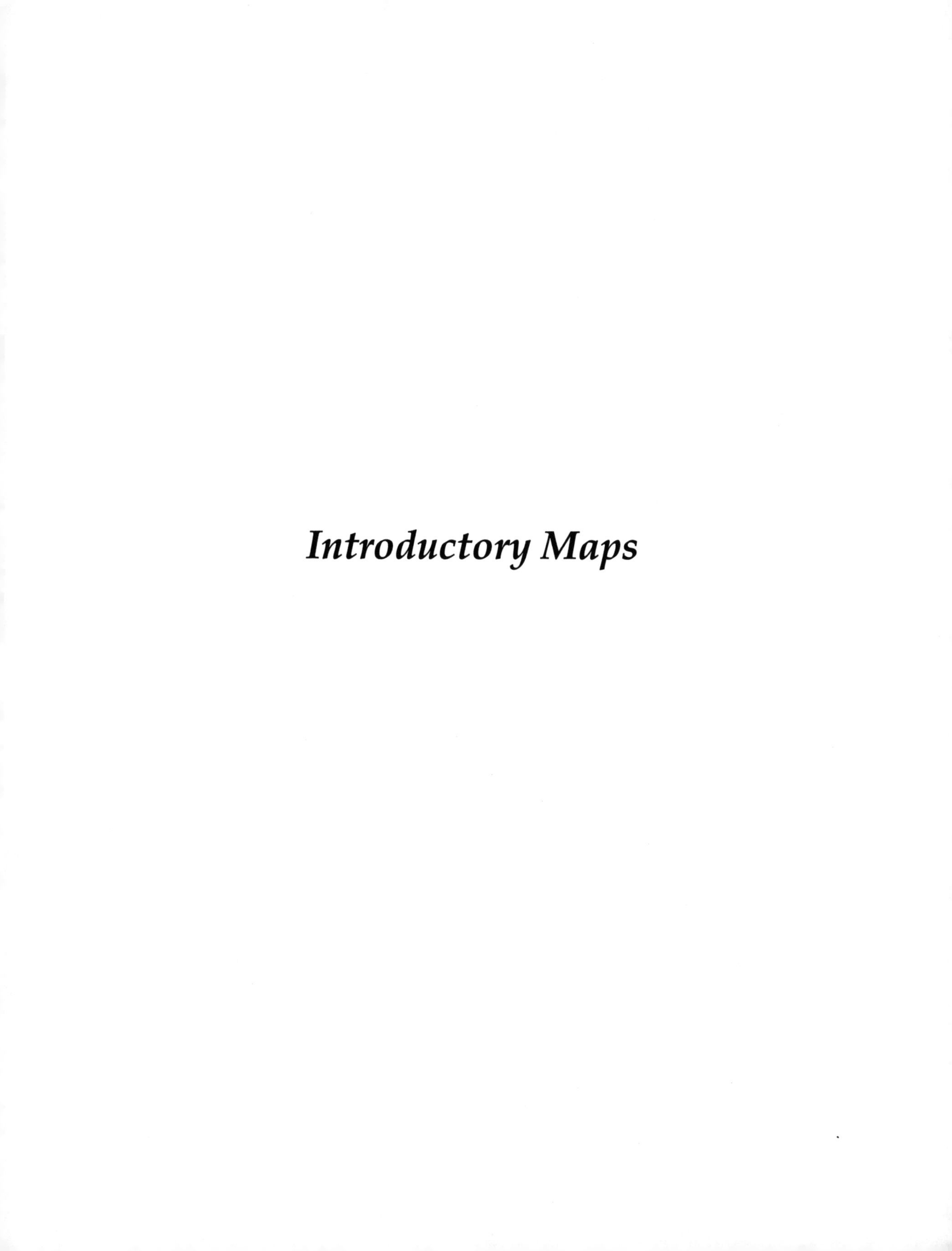

Introductory Maps

Map 1: Physical

The Balkan Peninsula (the "Balkans") often has been labeled "Southeastern Europe." It is a rugged, irregular, inverted triangle of land jutting southward from Europe into the eastern Mediterranean Sea, bounded on the west by the Adriatic, on the east by the Black, and on the southeast by the Aegean seas. Its northern land border is defined only partially by mountains. The Carpathian Mountains provide a limited boundary to parts of the north and northeast, while the Julian Alps delineate its extreme northwestern corner. Roughly 300 miles (480 kilometers) of open land carved by the Danube, Sava, and Drava rivers divide those two chains in the northwest, while, in the northeast, the plains and tablelands of the Danube and Prut rivers separate the Carpathians from the Black Sea by some 125 miles (200 kilometers). In the northwest, the Sava and Drava rivers often have been designated boundaries in the open plains. Similarly, the Prut River has been used as a Balkan boundary on the northeastern plain. Using the Drava and Prut rivers as part of the northern geographical boundary, the Balkan Peninsula encompasses some 276,700 square miles (716,650 square kilometers) of territory.

Crisscrossing mountain chains characterize the Balkans' terrain. The name "Balkan" derives from a colloquial Turkish word for a forested mountain and is now also the accepted name of a particular mountain chain south of the Danube (known in classical times as the Haimos [Hæmus]) stretching from the Black Sea for half the peninsula's east-west width. To its south stretch a densely grouped series of mountain ranges—the Rila (with the highest peak in the Balkans: 9,592 feet [2,926 meters]), the Rhodope, the Pindos, and the Taigetos—to the tip of the peninsula in the Peloponnese. The peninsula's west is dominated by the ruggedly karstic Dinaric and Albanian alps, which run parallel to the Adriatic coastline but spread extensively inland. Close to 70 percent of the peninsula is covered by mountains.

Except for the narrow coastal plains, river valleys constitute most of the lowlands. The largest is that of the Danube, which cuts a wide swath of plain and tableland into the heart of the peninsula between the Dinaric Alps and the Carpathian Mountains, narrowing at the so-called Iron Gates, where the river literally carves a gorge separating the Balkan and Carpathian mountains, before widening once again into a broad plain extending to the Black Sea. Others, such as the Drava, Sava, Morava, and Iskŭr river systems (important branches of the Danube watershed), the valleys of the Aliakmon, Vardar, Struma, Mesta, and Maritsa rivers (draining into the Aegean), and the Neretva, Drin, Shkumbin, and Vijosë river valleys (running to the Adriatic), provide a modicum of arable land in the mountainous interior and the only lines of natural overland communication.

Climatically, the Balkan Peninsula is not a unit. It enjoys a Mediterranean climate along most of its sea coasts and a continental one throughout its interior. Vegetation and land use vary with the natures of the dual climate. Along the coasts, the land mostly is rocky and denuded, supporting such crops as olives, grapes, figs, lemons, and oranges, and the herding of sheep and goats. In the interior, most of the mountains are forested; cereal crops predominate in the river valleys and lowlands; vineyards are found in some areas of the Danubian Plain, in the Maritsa River valley, and along the upper Sava; and livestock breeding mostly involves pigs and cows, though sheep and goats are fed on highland pastures. The line separating the two climate zones lies close to the coastline in most of the peninsula, since the mountains, which form the climatic border, push close to the seas almost everywhere.

Geopolitically, the peninsula is of historical strategic significance. It is a crossroads of three continents—Europe, Asia, and Africa—and its accessibility by both sea and land lays it open to political, military, and cultural human incursions and contentions from all directions. In the past, six foreign empires—the Persian, Roman, Byzantine, Ottoman, Habsburg Austrian, and Russian—sought to possess, whole or in part, the benefits offered by the peninsula's strategic location and natural resources with varying degrees of success.

Interspersed among the foreign imperial efforts were those of indigenous Balkan states. Because of its rugged geography and harsh interior climate, political life in the Balkans historically has been unstable. Small states, beginning with the classical Greek city-states, have been the rule because of the mountainous topography, which tend to separate human habitation in isolated river valleys and highland plateaus. This resulted in centuries of fierce competition for control of the geographically restricted available natural resources. Because Balkan states nearly always proved vulnerable to outside empires competing for sway in the region, those resources rarely benefited the inhabitants.

When Balkan states other than the Byzantine Empire managed to survive for any length of time, they did so mostly in the peninsula's interior, and the geographic division between coast and mountains had economic consequences. Often the coast, with its important seaports, was controlled by foreign states frequently at odds with native ones in the interior, thus effectively barring the latter from secure outlets to the seas. For such reasons, the economies of Balkan states primarily remained agricultural long into the 20th century.

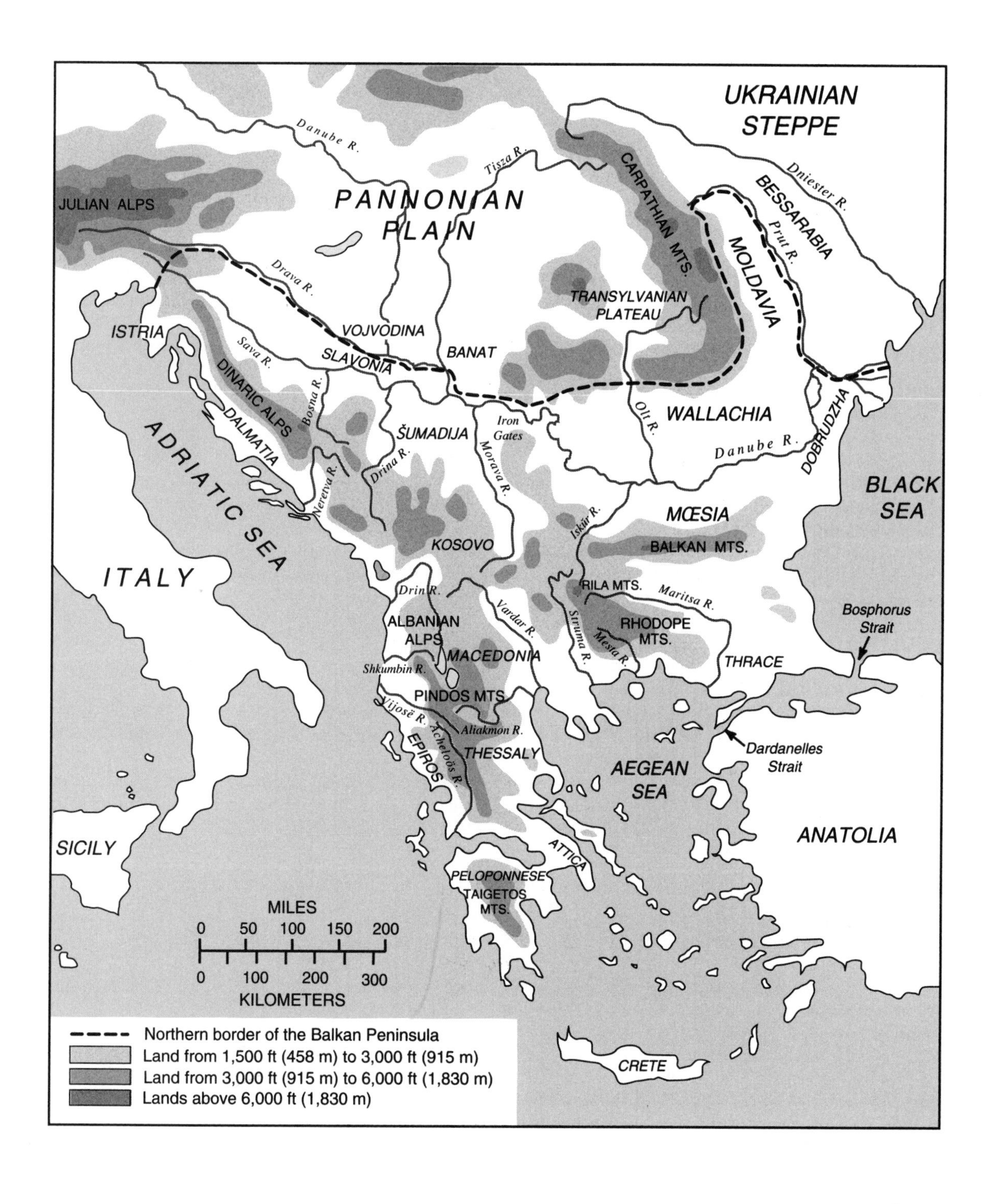

UKRAINIAN STEPPE
Danube R.
Tisza R.
Dniester R.
BESSARABIA
JULIAN ALPS
PANNONIAN PLAIN
CARPATHIAN MTS.
Prut R.
MOLDAVIA
Drava R.
TRANSYLVANIAN PLATEAU
ISTRIA
VOJVODINA
Sava R.
SLAVONIA
BANAT
DINARIC ALPS
Bosna R.
DALMATIA
Iron Gates
ŠUMADIJA
Olt R.
WALLACHIA
DOBRUDZHA
ADRIATIC SEA
Drina R.
Morava R.
Danube R.
Neretva R.
BLACK SEA
Iskăr R.
MŒSIA
KOSOVO
BALKAN MTS.
ITALY
Drin R.
RILA MTS.
Maritsa R.
Vardar R.
Struma R.
Bosphorus Strait
ALBANIAN ALPS
RHODOPE MTS.
Mesta R.
MACEDONIA
THRACE
Shkumbin R.
PINDOS MTS.
Vijosë R.
Aliakmon R.
Achelöos R.
EPIROS
THESSALY
Dardanelles Strait
AEGEAN SEA
ANATOLIA
SICILY
ATTICA
PELOPONNESE
TAIGETOS MTS.
MILES
0 50 100 150 200
0 100 200 300
KILOMETERS
Northern border of the Balkan Peninsula
Land from 1,500 ft (458 m) to 3,000 ft (915 m)
Land from 3,000 ft (915 m) to 6,000 ft (1,830 m)
Lands above 6,000 ft (1,830 m)
CRETE

Map 2: Political, 2001

ALBANIA

Capital: Tiranë
Area in sq. mi. (sq. km.): 11,097 (28,489)
Population: 3,334,000
Ethnicity (%): Albanian (90); Greek, Vlah, Gypsy, Bulgarian (10)
Languages: Albanian; Greek
Religions (%): Muslim (70); Albanian Orthodox (20); Roman Catholic (10)
Type of Government: Parliamentary democracy

BOSNIA-HERCEGOVINA

Capital: Sarajevo
Area in sq. mi. (sq. km.): 19,776 (51,233)
Population: 4,618,800 (prior to 1992-95 war)
Ethnicity (%): Bosniak/Muslim (44); Serbian (31); Croatian (17); other (8)
Language: Serbian; Croatian
Religions (%): Muslim (40); Orthodox (31); Roman Catholic (15); Protestant (4); other (10)
Type of Government: Emerging parliamentary democracy

BULGARIA

Capital: Sofia
Area in sq. mi. (sq. km.): 42,855 (110,994)
Population: 8,832,000
Ethnicity (%): Bulgarian (85); Turkish (9); Macedonian, Gypsy, other (6)
Languages: Bulgarian; Turkish; other
Religions (%): Bulgarian Orthodox (85); Muslim (13); Roman Catholic, Jewish, Protestant, other (2)
Type of Government: Republic

CROATIA

Capital: Zagreb
Area in sq. mi. (sq. km.): 21,824 (56,538)
Population: 4,694,400
Ethnicity (%): Croatian (78); Serbian (12); Hungarian, Slovenian, other (10)
Languages: Croatian; Serbian; other
Religions (%): Roman Catholic (77); Orthodox (11); Muslim (1); Protestant (1); other (10)
Type of Government: Parliamentary democracy

GREECE

Capital: Athens
Area in sq. mi. (sq. km.): 50,962 (131,990)
Population: 10,010,000
Ethnicity (%): Greek (93); Macedonian (2); Turkish (1); other (4)
Languages: Greek
Religions (%): Greek Orthodox (97), Muslim (1); other (2)
Type of Government: Republic

MACEDONIA

Capital: Skopje
Area in sq. mi. (sq. km.): 9,778 (25,333)
Population: 2,194,000
Ethnicity (%): Macedonian (67); Albanian (21); Turkish (4); Serbian (2); other (6)
Languages: Macedonian; Albanian; Turkish; Serbian; other
Religions (%): Orthodox (59); Muslim (26); Roman Catholic (4); Protestant (1); other (10)
Type of Government: Parliamentary democracy

ROMANIA

Capital: Bucharest
Area in sq. mi. (sq. km.): 91,699 (237,499)
Population: 23,172,000
Ethnicity (%): Romanian (89); Hungarian (9); German, Turkish, Russian, Ukrainian, Gypsy, other (2)
Languages: Romanian; Hungarian; German
Religions (%): Romanian Orthodox (70); Roman Catholic (6); Protestant, Jewish, other (6)
Type of Government: Democratic republic

SLOVENIA

Capital: Ljubljana
Area in sq. mi. (sq. km.): 7,834 (20,296)
Population: 1,967,700
Ethnicity (%): Slovenian (91); Croatian (3); Serbian (2); other (4)
Languages: Slovenian; Croatian; other
Religions (%): Roman Catholic (96); Muslim (1); other (3)
Type of Government: Parliamentary democracy

YUGOSLAVIA
(FEDERATION OF SERBIA AND MONTENEGRO)

Capital: Belgrade
Area in sq. mi. (sq. km.): 39,507 (102,350)
Population: 10,700,000
Ethnicity (%): Serbian (63); Albanian (14); Montenegrin (6); Hungarian (4); other (13)
Languages: Serbian; Albanian
Religions (%): Orthodox (65); Muslim (19); Roman Catholic (4); Protestant (1); other (11)
Type of Government: Federated democratic republic

UKRAINE
Danube R.
Vienna
AUSTRIA
Budapest
MOLDOVA
HUNGARY
SLOVENIA
Venice
Ljubljana
Zagreb
CROATIA
ROMANIA
BOSNIA-
HERCEGOVINA
Belgrade
Bucharest
YUGOSLAVIA
Sarajevo
Danube R.
SERBIA
MONTENEGRO
Podgorica
Sofia
ITALY
ALBANIA
BULGARIA
Skopje
MACEDONIA
Tiranë
Istanbul
GREECE
TURKEY
Athens
MILES
0 50 100 150 200
0 100 200 300
KILOMETERS

Map 3: Natural Resources

Albania is divided into two zones: Coastal plains and extensive interior highlands (Albanian Alps). Most rivers—the Drin, Shkumbin, and Vijosë—run from the highlands to the Aegean Sea. Cereal production and some Mediterranean-type agriculture are conducted on the coastal plains, while forestry and livestock pasturing predominate in the highlands. Chromium and copper represent the most important mineral resources.

Bosnia-Hercegovina lies within the Dinaric Alps. Most available arable land is in the Neretva, Bosna, and Drina river valleys or in scattered mountain basins and plateaus. Extensive forests make timber products important. Livestock is herded in upland pastures. Lignite, iron, and manganese are mined in Bosnia, while bauxite and lignite are worked in Hercegovina. In former times, Bosnia was important in gold and silver mining.

Bulgaria's borders encompass the Balkan, Rila, and Rhodope mountains and an extensive network of river systems, which generally flow north into the Danube (the Iskŭr River) or south into the Aegean (the Maritsa, Struma, and Mesta rivers). Its Thracian and Danubian plains are excellent for cereal and fruit cultivation. A uniquely important crop is attar of roses—a crucial and expensive ingredient in most top-of-the-line perfumes. Pigs, sheep, cows, and goats are herded. Coal (both black and brown), iron, copper, zinc, and lead are important mineral resources.

Croatia consists of three regions: Mountainous Croatia Proper, from which stretch the two horns of Slavonia (the lowlands between the Sava and Drava rivers), and Dalmatia (the Adriatic coastline and Dinaric highlands). The Sava and Drava are the primary river systems. Croatia Proper and Dalmatia are highly forested. Livestock is herded on upland pastures, while grains are sown in depressions and valleys. Mediterranean-type cultivation proliferates in Dalmatia. In Slavonia, cereal and fruit crops predominate. Mineral resources are limited, consisting of small pockets of iron, natural gas, oil, and bauxite.

Greece's mountains cover 80 percent of its mainland, rendering less than a third of its land suitable for cultivation. The Pindos Mountains run the north-south length of the state, breaking at the Gulf of Corinth, but continue in the Peloponnese as the Taigetos range. The mountainous interior is linked to surrounding seas by the Aliakmon and Acheloös river systems. Most cultivation is restricted to the narrow coastlines, where typical Mediterranean-type crops are produced. Cereals are grown on scattered upland plateaus, the Thessalian Plain, and the Macedonian-Thracian coastal plain. In the mountains, Mediterranean-type scrub and pasture predominate. Although a variety of ores are present—bauxite, iron, copper, lead, zinc, and silver—they exist in small amounts of little economic benefit. Greece mostly depends on maritime trade.

Macedonia is a mountainous region where the southern Dinaric and eastern Albanian alps meet the Pindos Mountains' northern projections. The Vardar River system, which bisects the state from north to south, provides it with its principal lowlands. On these are grown cereals, tobacco, cotton, and some fruits, as well as wine grape-producing vines. Close to half of the state is heavily forested but there is some upland pasture for sheep and goats, along with cultivation in valleys and depressions. Mineral resources include small deposits of zinc, lead, iron, chrome, and manganese. In times past, gold was mined in its eastern regions.

Romania is divided topographically into two basic arable zones—the plains of the Danube and Prut rivers and the Transylvanian Plateau—by the boomerang-shaped, territorially extensive Carpathian Mountains. All of the river systems draining the Wallachian Plain and Moldavian tablelands flow from the Carpathians to the Danube, while those of the Transylvanian Plateau mostly run westward, emptying into the Tisza River on the Pannonian Plain. The Carpathians and the Transylvanian highlands are thickly forested, providing pasture for sheep and goats and some cultivation. The Wallachian and Moldavian lowlands, as well as Transylvania's western edge, are cultivated with grains, flax, hemp, tobacco, and wine grape-producing vines. The state possesses the richest and most diverse mineral resources in the Balkans. Europe's largest continental oil fields lie in the southern Carpathian foothills and Transylvania is rich in natural gas. Large deposits of salt, lignite, black and brown coal, copper, and iron, supplemented by zinc, manganese, silver, gold, and mercury, are mined in the Carpathians and Transylvanian Plateau.

Slovenia lies within the terminal ranges of the Julian Alps. Most land not covered by mountains consists of highly forested foothills and depressions, cut through by the upper courses of the Sava and Drava rivers. The largest lowland is the Drava Basin, where grains and some fruits and vines are grown. Elsewhere, cultivation is undertaken in numerous depressions and valleys. Livestock pasturing and lumbering are widespread. Mineral resources include oil and natural gas, brown coal, lignite, zinc, lead, and mercury.

Yugoslavia (Serbia and Montenegro) occupies the north-central Balkan regions. The Dinaric Alps dominate the landscape in the south and southwest, and Montenegro lies almost completely within their folds. The state possesses only a short Adriatic coastline in Montenegro, although the Morava River provides a relatively direct route to the Aegean Sea by way of the Vardar River valley. South of Belgrade lies a somewhat forested hill country (Šumadija), drained by the extensive Morava River system. The full range of continental crops are cultivated in the Danubian lowlands and Šumadija. The forested mountain regions offer pasturing and some grain cultivation. In areas of mountainous, barren Montenegro, Mediterranean-type cultivation is undertaken. Montenegro can claim only bauxite as significant, whereas Serbia is endowed with brown coal, lead, and zinc, along with lesser amounts of black coal and copper.

Key:

Ag	Silver	Bn.C.	Brown Coal	Fe	Iron	Ni	Nickel
Al	Aluminum	Bit.	Bitumen	Hg	Mercury	Pb	Lead
Au	Gold	Bx	Bauxite	Lig.	Lignite	Pet.	Petroleum
B.C.	Black Coal	Cr	Chrome	Mn	Manganese	Sb	Antimony
		Cu	Copper	N.G.	Natural Gas	Zn	Zinc

Map 4: Demographic

The Balkan Peninsula's population of approximately 69.3 million people (excluding European Turkey) essentially is composed of three primary groupings: Historically ancient peoples, South Slavs, and Turks. There exist a smattering of numerically smaller groups.

Ancient peoples (those who trace their ethnic ancestors back at least to classical antiquity) account for some 50 percent (roughly 35 million). The most familiar are the Greeks, who generally occupy their ancestral territories, despite the 6th- and 7th-century Slavic invasions and settlements of their mainland possessions. Only a long process of military reconquest by the Greek-speaking Byzantine Empire permitted the Greeks to regain their ancient homeland's interior, but pockets of Slavic speakers survived as far south as the Peloponnese and in Macedonia. Though some argue that the Slavic incursions into Greek-inhabited regions diluted the modern Greeks' direct genetic link to their ancestors, language (and the self-identity that it bestows), not DNA, is the fundamental measure of ethnicity.

Albanians speak a unique language thought to have descended from ancient Illyrian, making them contemporaries of the Western European Basques. In antiquity the Illyrians occupied a large swath of the western Balkans. Waves of Roman, Goth, Avar, and Slav invaders pushed the Illyrians into the mountainous regions inhabited by today's Albanians, where they evolved as a mostly tribalized, pastoral society divided into two distinct subgroups identified by dialect: The northern Ghegs and the southern Tosks.

The Romanians speak a Latin-based language that, in Romanian national thinking, derives from the Roman occupation of ancient Dacia during the 2nd and 3rd centuries. When the Romans withdrew, the Latinized Daks remained behind in the Carpathian Mountains, surviving successive waves of Germanic, Slavic, and Turkic invaders and reemerging in the 13th century ethnically unscathed to occupy their present territories. This claim is contested by many non-Romanians, who suggest that the Romanians originated as pastoral, Latin-speaking Vlahs south of the Danube who migrated into present-day Romania some time after the late 9th century. The name of the Romanian region Wallachia means "Land of the Vlahs." The Vlahs may have descended from Latinized Thracians, who inhabited the Thracian Plain (to which they lent their name). A pastoral, seminomadic tribal people, by the 13th century they acquired the name "Vlah." Today they constitute an ethnic minority in all central and southern Balkan states, and their number is dwindling (less than 100,000) because of continual assimilation into dominant ethnic groups.

South Slavs are the second major Balkan ethnic component, numbering some 29 million people (over 41 percent), divided today among seven major groups: Bosnians, Bulgarians, Croats, Macedonians, Montenegrins, Serbs, and Slovenes. South Slavs form one of the three primary branches of European Slavic speakers, the others being the West and East Slavs. Their common ancestors entered Eastern Europe during the 5th through 7th centuries from a homeland that lay in the vicinity of the Pripet Marshes, which straddle the border between today's Ukraine and Belarus. They came as part of the lengthy human migratory process commonly called the "Barbarian Invasions of Europe." Initially, all Slavs spoke dialects of a common language, but tribal migrations in three generally different directions, the passage of time, and later settlements of non-Slavic peoples in Central Eastern Europe resulted in the formation of three distinct subgroups of Slavic speakers. The South Slavic groups moved south and southwest from their Pripet homeland, eventually entering the Byzantine-controlled Balkans, along with invading Turkic Avars, during the second half of the 6th century and establishing new, permanent settlements. Today, their descendants solidly inhabit the northwestern, central, and southeastern Balkan regions.

Turks are the third major ethnic component of the Balkans' population. Though today numerically small—a little over 1 million people (about 2 percent)—they have great significance in Balkan history. Huns, Avars, and related tribes and allies swept through the Balkans in the 5th through 7th centuries, and among them were the Bulgars, who arrived in the late 7th century and established a state. Unlike the Balkan settlements of the Avars, which proved transitory, the Bulgar state persisted, and by the 9th century, the Bulgars challenged the Byzantine Empire for Balkan political hegemony. By that time, they also rapidly were assimilating into their largely Slavic-speaking subject population. The conversion of the Turkic Bulgars to Orthodox Christianity at mid-century opened the gate to their total Slavic assimilation, and most traces of their Turkic origins disappeared, except for their name—the Bulgars become Slavic Bulgarians.

Oğuz, Pecheneg, and Cuman Turkic tribes appeared in the Balkans between the 9th and 11th centuries. Most of them suffered an ethnic fate similar to the Bulgars and left little lasting impression, though it is possible that some Turks living today in the eastern Balkans and the Gagauz Turks of Bessarabia (now known as Moldova) are direct ethnic descendants of those medieval Turkic interlopers. Additionally, the Ottoman Turks' five-century rule over most of the Balkans established numerous scattered enclaves of Turkish-speaking groups throughout the peninsula's southern regions, with a heavy concentration in Thrace.

The remaining 7 percent of the peninsula's inhabitants (slightly over 4 million people) usually are lumped together under the category of "Other" in demographic statistics and include Jews, Gypsies, Hungarians, Czechs, Slovaks, Germans, Italians, Mongol-Tatars, Russians, Ukrainians, and Ruthenians. The striking ethnic diversity of the "Other" groups lends the Balkans one of its most distinctive characteristics.

Danube R.
UKRAINE
AUSTRIA
HUNGARY
MOLDOVA
SLOVENIA
Slovenes
CROATIA
BOSNIA-HERCEGOVINA
Bosnians
SERBIA
YUGOSLAVIA
Serbs
Montenegrins
MONTENEGRO
ROMANIA
Romanians
Székelys
Danube R.
BULGARIA
Bulgarians
Turks
MACEDONIA
Macedonians
Albanians
ALBANIA
GREECE
Greeks
ITALY
TURKEY
MILES
0 50 100 150 200
0 100 200 300
KILOMETERS
South Slavs
Major Ethnic Minorities:
AL Albanians
BG Bulgarians
CR Croats
GK Greeks
GR Germans
HU Hungarians
MA Macedonians
MN Montenegrins
SB Serbs
TK Turks

Map 5: Cultural

To make sense of Balkan history, one must understand the cultural forces that historically operated in the region. Human culture usually operates on two levels—the civilizational and the ethnonational. Civilization is a complex culture shared by a network of ethnonationally identified peoples spread over a large geographic area that demonstrates highly developed, sophisticated institutions, widespread urbanization, a written language, and both division of labor and social differentiation. Ethnonationality is a group identity usually conveyed by a specific language. Every civilized society incorporates a number of ethnonational societies that are unified by a common religious belief, philosophy, or both.

The Balkans historically have witnessed the interactions of three civilizations: The Orthodox Eastern European, the Western European, and the Islamic. The Eastern and Western European civilizations share in common important cultural traditions. Classical Greco-Roman, Hellenic civilization, "barbarian" (non-Roman) ancestors, and Christianity, which culturally define "Europe." The cultural differences inherent in Greco-Roman traditions—Greek "eastern" and Latin (Roman) "western"—account for their developing as two separate civilized societies. The "east" espoused the classical Greeks' universal, mystical, idealist sense of reality; the "west" embraced the Romans' practical, legalistic, pragmatic approach. Those differences initially were institutionalized in the forms of Christianity each developed—Orthodoxy ("east") and Catholicism/Protestantism ("west"). The two "Europes" actually are cultural twins, sharing the same genetic makeup but differing in character. Islamic civilization is their cultural cousin, sharing a good deal of their Judeo-Christian and Hellenic traditions but embracing crucially different Arabic and Mesopotamian characteristics. Islam views itself as the divinely ordained corrective for deficiencies that crept into Judaism and Christianity.

As an important territorial component of the East Roman/Byzantine Empire, which provided the sociopolitical environment for Orthodox Eastern Europe's creation, the Balkans can be considered Orthodox Eastern European civilization's homeland. Only its northwestern corner (Slovenia, Croatia, part of Bosnia, and northern Albania) lay outside that development. For most of the Balkans, both the Western European and Islamic civilizations were "foreign imports." The Islamic one was imposed by force, and the Western European one entered the Orthodox Eastern European Balkans through "cultural botany"—grafting Western cultural elements onto a closely related Greco-Roman cultural trunk (beginning with the adoption of Western-style nationalism).

To simplify our understanding of civilizational interactions, we might borrow the geological analogy of continental plate tectonics. Every civilization can be said to hold sway over a large, geographically defined core area in which the fundamental worldview binding together its constituent societies has undergone native, organic development. Each core area might thus be viewed as a large plate superimposed on the map of the world. Focusing on the Balkans, we find the Orthodox East European plate generally sandwiched between the Western European (to its northwest) and the Islamic (to its south). Their points of contact form cultural fault lines just as tectonic plates form seismographic faults.

Human fault lines cannot be delineated neatly on a map. After centuries of human interaction, ethnonational societies have penetrated into regions dominated by others. Human cultural faults, therefore, are depicted on our map as bands of green whose widths vary by location and history. These bands represent lines along which occur the most dramatic human disturbances, caused by friction among the differing civilizational plates.

Historically, the cultural fault line dividing the Western and Eastern European civilizations in the Balkans runs from Transylvania in Romania, through Serbia's (Yugoslavia's) Vojvodina province, the Slavonian border region separating Croatia and Serbia, all of Bosnia-Hercegovina, to the Dalmatian-Montenegrin border and northern Albania along the Adriatic Sea. A second fault line separates the Eastern European and the Islamic civilizations. Although seemingly short—it parallels the border of Turkey with Bulgaria and Greece—Islamic Ottoman conquest and centuries of Ottoman rule created an extensive band resembling a long, somewhat scythe-shaped swath cutting northwestward into the peninsula through Bulgaria, northern Greece, Macedonia, Albania, and Kosovo, eventually intersecting the East-West European fault in Bosnia-Hercegovina and northern Albania. Therefore, the faults of the three Balkan civilizations all converge in Bosnia-Hercegovina and northern Albania.

Human frictions along and within the Balkans' civilizational faults have been long-lived, with frequent flare-ups. Since the 9th century, Transylvania, Banat, Vojvodina, Slavonia, Bosnia-Hercegovina, Greece, and northern Albania frequently have witnessed human-cultural seismic eruptions between Orthodox East European and Western European societies. They were most intense during and following the medieval Crusades and during the highly nationalist late 19th and 20th centuries. The East European-Islamic fault was formed in the 14th century and was pushed steadily northward by the Ottoman Turks until the 17th century. The fault was pressed to its present location in the early 20th century. Every region of the Balkans has served as a cultural flashpoint at least once, with Bosnia-Hercegovina, Macedonia, Slavonia, Kosovo, and Bulgaria proving the most recent.

Within the Balkans, the major ethnonational groups that belong to Orthodox Eastern European civilization are some Albanians, Bulgarians, Greeks, Macedonians, Montenegrins, Romanians, and Serbs, as well as some Gypsies and Vlahs. Societies of Western European civilization include a small number of Albanians, Croats, and Slovenes. Balkan Islamic societies are most Albanians, Bosniaks (Bosnian Muslims), Turks, Pomaks (in Bulgaria, Greece, Macedonia, and Serbia), and some Gypsies and Vlahs.

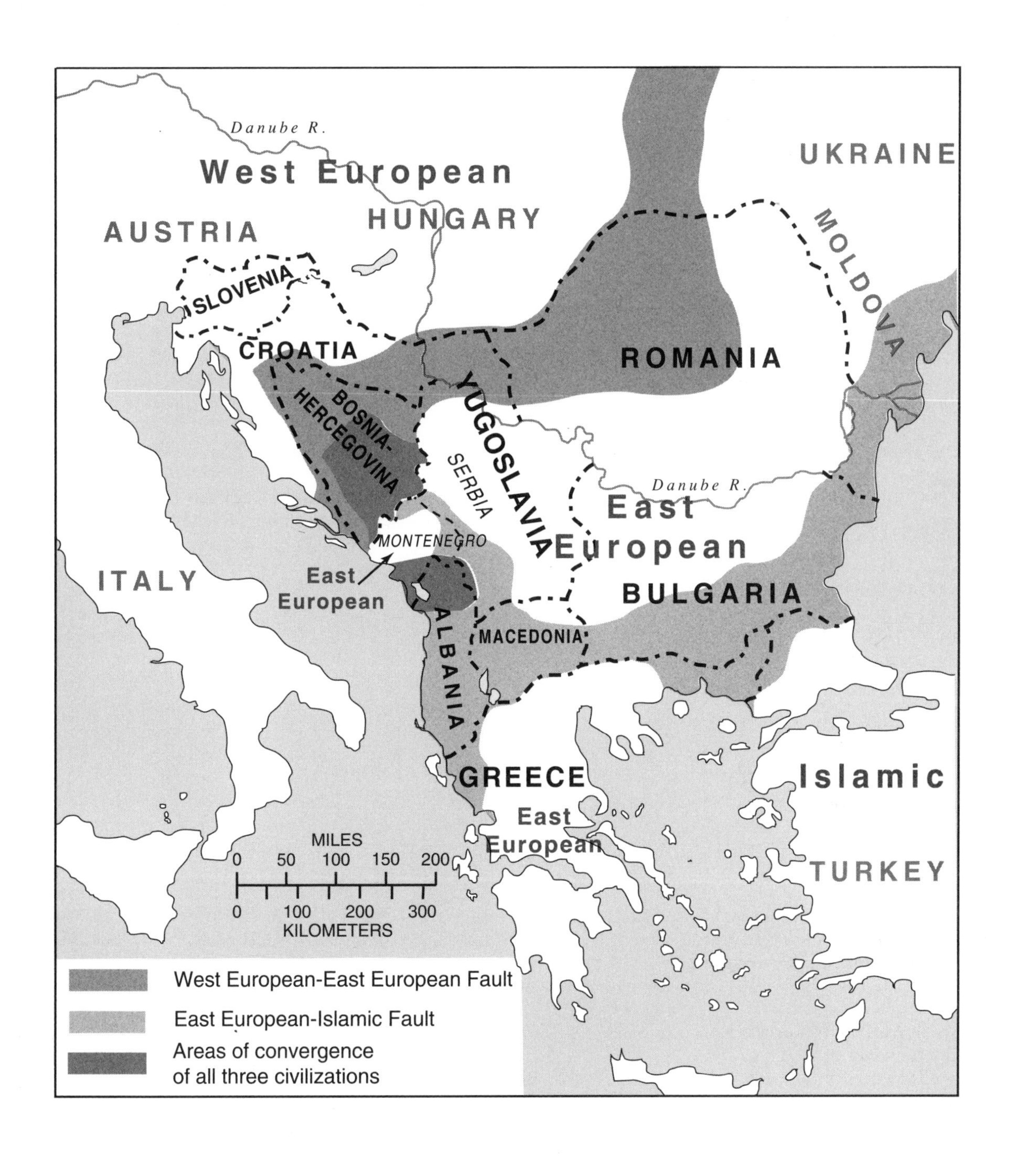

Danube R.
West European
UKRAINE
AUSTRIA
HUNGARY
MOLDOVA
SLOVENIA
CROATIA
ROMANIA
BOSNIA-HERCEGOVINA
YUGOSLAVIA
SERBIA
Danube R.
East European
MONTENEGRO
East European
ITALY
BULGARIA
MACEDONIA
ALBANIA
GREECE
Islamic
East European
TURKEY
MILES
0 50 100 150 200
0 100 200 300
KILOMETERS
West European-East European Fault
East European-Islamic Fault
Areas of convergence of all three civilizations

Era of Byzantine Hegemony
600-1355

Map 6: The East Roman Balkans, Late 6th Century

Until Roman Emperor Constantine I the Great (306-37) established his imperial capital at Constantinople (Byzantion), the Balkans were little more than a transit zone for west-east Roman military movements and an economic backwater. Only its ports of Durrës (Dyrrachion), Thessaloniki, and Byzantion were held in high economic account. Overland travel was restricted to narrow coastal lowlands and through a limited number of mountain passes by means of a simple road network by Roman standards, consisting of the *Via Ignatia* (the most direct route linking the Roman west and east), the Belgrade (Singidunum)-Byzantion Diagonal Highway *(Via Militaris)*, the Morava-Vardar Highway linking those two through Macedonia, and the Mœsian Highway along the Danubian Plain. The limited nature of the road network constrained military, economic, and personal movement.

The situation changed somewhat with the founding of Constantinople (the City of Constantine), since populations living close to the major highways enjoyed a period of commercial prosperity and expanded urbanization. Most of the empire's Balkan population, however, remained rural, working land divided among free small holdings, free village holdings, and large estates owned by magnates. The exact proportion of each in the landholding scheme remains unclear, but most likely the share of great landed estates generally rose at the expense of the others during the 4th through 6th centuries. State taxes escalated to pay for the military expenses of the defensive army established by Constantine I. The empire's borders were manned by frontier troops whose task was to hinder invaders until the emperor's mobile elite force arrived to deliver the decisive blow. The elite force (usually barbarian mercenaries) was organized into formal military units and stationed strategically in the interior.

The Balkans' mountains provided military defense positions protecting the interior against attacks from the north, since the primary invasion routes could be blocked effectively by fortifying the main highways' passes. Lacking fortifications, ambushes could suffice. A military drawback to the mountain defenses was the inability to fortify all of the numerous passes cut by lesser tributary rivers scattered throughout all of the ranges. An enemy with knowledge of them could outflank the principal fortifications guarding the main routes.

Despite the increased prosperity of the Balkan provinces following the 4th century, their intrinsic value to the empire never equaled those in the east, nor did they match Italy in terms of imperial prestige. If troops were needed in the east against Persia, the peninsula's elite forces were dispatched. When Emperor Justinian I the Great (527-65) set out to reconquer North Africa and Italy in the west from the Goths, his professional forces mostly were drawn from the Balkans. In the elite forces' absence, the peninsula lay vulnerable to attacks from the north. Risks against such assaults were taken in the Balkans because the emperors felt that they could depend on the mountains and frontier forces to slow an enemy incursion until the elite units were freed from duties elsewhere and returned to deal with the situation. If in-depth frontier defense proved insufficient and an enemy broke through before the elite forces returned, the emperors could depend on the capital to play the delaying role of last resort.

Constantinople, located at the economic and military crossroads of Europe and Asia, was the largest and strongest fortress-city in Europe and virtually impregnable. Situated on an easily defensible triangular bit of land on the European shore of the Bosphorus Strait, it was surrounded by a sea wall on its southern and northern sides and a triple land wall on its western. For a thousand years the land walls of Constantinople often saved the empire from defeat or even utter destruction by its enemies.

Constantinople's walls protected more than the empire—they guarded the womb of an emerging Orthodox European civilization. The city gave concrete expression to Constantine's policy of reconstituting the Roman Empire on a Christian moral basis and symbolized a new imperial political ideology. The Christian state church represented the true believers' temporal community whose borders were synonymous with those of the Roman Empire. State and church were united in an indissoluble partnership that was believed to reflect a divinely ordained world order. Administering that order was the Roman emperor—God's viceroy on earth and "Thirteenth Apostle of Christ"—who ruled as a Christian autocrat from his capital at Constantinople.

Constantine laid the groundwork for the new order's emergence and Justinian I assured its success. Justinian spent his reign relentlessly pursuing imperialist goals to the point of driving the empire into near bankruptcy by the time of his death. He ordered the great collection and codification of Roman law that entrenched Christian precepts and eventually came to serve as the legal foundation for state government in much of both Eastern and Western Europe until the late 18th century. He sponsored expensive building campaigns to expand military defenses and, more importantly, to create a new form of Christian church architecture intended to express the power and glory of the partnership between Christianity and the Roman imperial office, epitomized in his magnificent imperial cathedral in Constantinople, Hagia Sophia (Holy Wisdom). Between 535 and 554 he waged almost constant warfare in the west. At its successful conclusion, his empire encompassed the western lands of Italy, Sicily, most of North Africa, and Spain's southern regions. Administrative and religious unity of East and West was reestablished and, over it all, Justinian briefly reigned supreme from Constantinople.

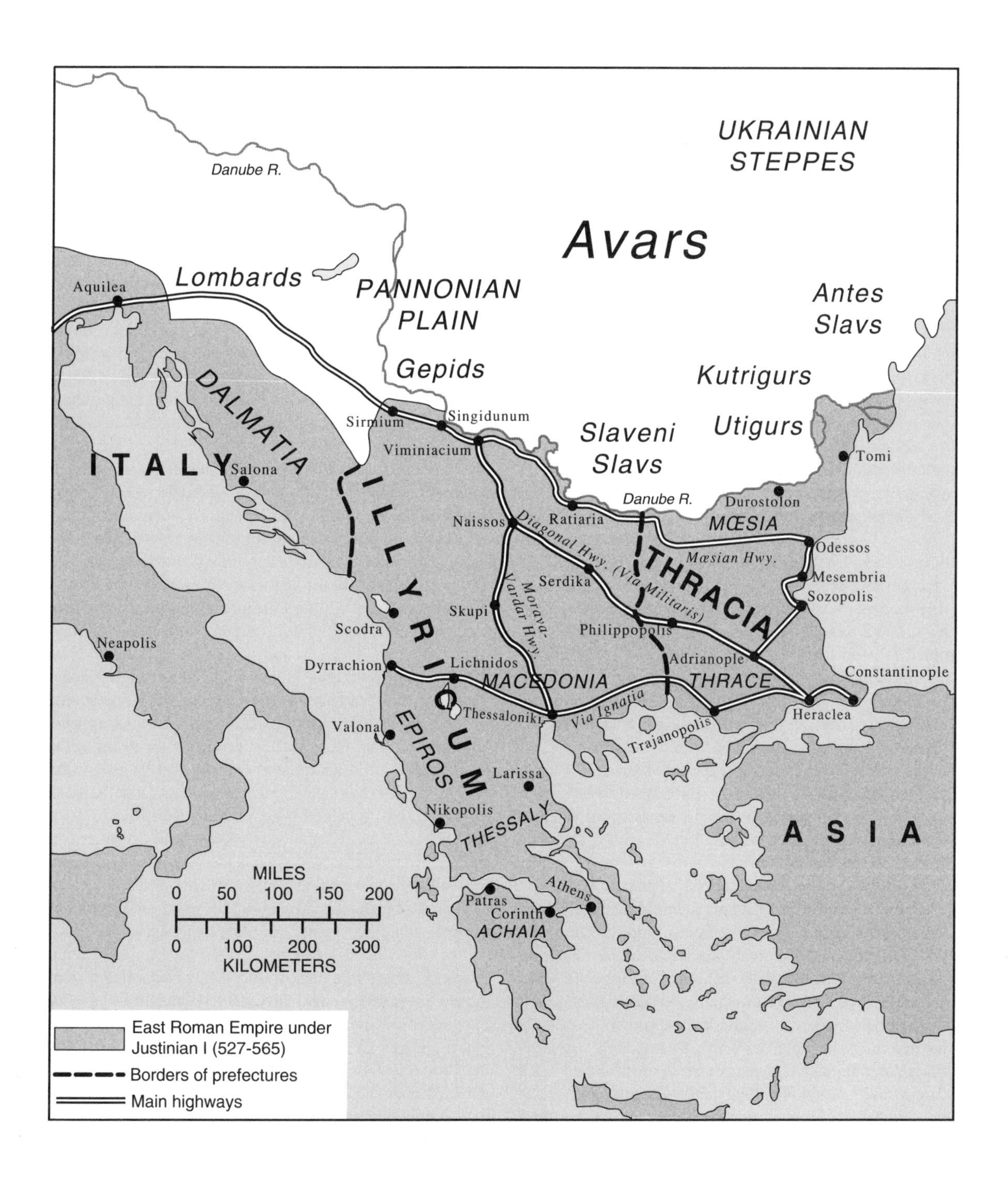

UKRAINIAN STEPPES
Danube R.
Avars
Lombards
PANNONIAN PLAIN
Antes Slavs
Aquilea
Gepids
Kutrigurs
DALMATIA
Sirmium
Singidunum
Slaveni Slavs
Utigurs
Viminiacium
Tomi
ITALY
Salona
Danube R.
Durostolon
Ratiaria
MŒSIA
Naissos
Diagonal Hwy. (Via Militaris)
Odessos
Mœsian Hwy.
THRACIA
Serdika
Mesembria
Sozopolis
Skupi
Morava-Vardar Hwy.
ILLYRICUM
Scodra
Philippopolis
Neapolis
Adrianople
Lichnidos
Dyrrachion
Constantinople
MACEDONIA
THRACE
Via Ignatia
Thessaloniki
Heraclea
Valona
EPIROS
Trajanopolis
Larissa
Nikopolis
THESSALY
ASIA
MILES
0 50 100 150 200
0 100 200 300
KILOMETERS
Athens
Patras
Corinth
ACHAIA
East Roman Empire under Justinian I (527-565)
Borders of prefectures
Main highways

Map 7: Avar, Slav, and Bulgar Invasions, 7th Century

In the early 6th century migrating Slav tribes arrived on Byzantium's (East Rome's) Balkan borders. Loosely divided into two different but related groups, Antes and Slaveni, they lacked any sophisticated political organization, essentially were disunited, and posed little more than a nuisance to the empire. That situation changed with the arrival of the mounted Turkic Avars in the 550s.

By the end of the 560s the Avars created a large confederated tribal state *(kaghanate)* north of the Balkans in Pannonia and conquered the Slav tribes north of the Danube. In the early 580s they turned the empire's Danube defenses in Mœsia and, with the Balkans' elite forces absent in the east fighting Persia, freely swept through the peninsula's heart to Thrace and Constantinople's environs. Large numbers of the formerly disorganized Slavs pushed south into the peninsula, either fleeing from the Avars or serving as their infantry allies, and it became apparent that they sought new territories to settle and not merely plunder. The Slavs' disunity and lack of state structure made them difficult for the Byzantines to deal with: They had no sufficiently powerful tribal leaders who could sign binding treaties or be bribed or subsidized effectively. The Slavs' primitiveness actually permitted them to settle and exist in harsh environments that more sophisticated populations avoided.

Wars with Persia in the east continually hindered Byzantium's efforts to thwart the Avar and Slav incursions, and bribing the Avar leadership merely slowed but never stopped their depredations. Emperor Maurice (582-602) briefly halted Avar and Slav inroads (602), but when he was overthrown by his mutinous troops, the Persians renewed the war, which dragged on for decades and drained the Balkans of elite forces. Thus, the former advantage over the Avars and Slavs was lost. In 626 the Persians launched a concerted assault on Constantinople through Anatolia in an attempt to crush the stricken Eastern Empire. They were joined by a massed Avar and Slav force, which threatened the capital from the European side. Avar assaults on the city's land walls were smashed and attempts to ferry Persian troops across the Bosphorus in small Slav boats were defeated by the imperial navy using Greek fire. Immediately thereafter, Avar operations disintegrated and the horde scattered northward in confusion, compelling the Persians to retreat.

The Avar *kaghanate* never fully recovered from the fiasco before Constantinople, and the Avar confederation slowly disintegrated. The West Slav tribes living in the north of the *kaghanate* rebelled against Avar control in the 620s and a number of allied Turkic steppe tribes broke away. So too did the South Slav tribes who, as a consequence of their involvement in Avar activities, now lay scattered in settlements spread throughout the Balkan Peninsula from the Danube to the Peloponnese.

Two related but different mixed Iranian-Slavic tribes broke from Avar authority and pushed into the northwestern and north-central Balkans—the Croats and Serbs. They fell on the Avars in the region and drove them into Pannonia, after which they brought the assorted local Slaveni Slavs under their control. Tradition credits Emperor Herakleios (610-41) with encouraging and blessing their actions by rewarding them with legal rights to the lands that they occupied, but it is unlikely that imperial policy played any direct role in the events. The Croats and Serbs established two loosely structured tribal confederations composed of the already settled Slavs. Both swiftly lost their Iranian ethnic characteristics and were assimilated into the Slavic culture of their more numerous subjects.

A political void existed in the Balkans during the second half of the 7th century following the successive blows of Avar depredations; Slav settlement; and imperial military, administrative, and economic contraction resulting from the empire's losses to the Arabs in the east. Into that void rode the Turkic Bulgars in the late 670s.

Bulgar steppe tribes previously tied to the Avars broke away after the defeat before Constantinople and, under Han (ruler) Kubrat (605-65), established a confederation of Great Bulgaria in the southern Ukraine. Kubrat then forged friendly relations with Byzantium. After Great Bulgaria was shattered by the Khazars in the early 640s, Kubrat's son Asparuh (d. 701) led the main branch of Bulgars southwestward to the Danube delta and subdued the Slavs and Avars on the Wallachian Plain. A separate Bulgar rebellion against the Avars in the late 670s was led by the chieftain Kuber (ca. 675-ca. 688), who headed an ethnically mixed "tribe" of Avar war prisoners in Slavonia. Kuber led his force south into the central Balkans and settled in northern Macedonia. Though his followers were Bulgars, Thracians, Illyrians, and Franks, the imperial authorities collectively identified them as "Bulgars."

In 679 Asparuh's Bulgars crossed the Danube and defeated the Byzantine border forces, winning a treaty from the empire in 681 recognizing a Bulgar state in former Roman territories south of the Danube. The Bulgar-controlled plains north of the river were joined to the lands ceded by the Byzantines, in which the Bulgars successfully exerted their political and military authority over the Slavic tribes that already inhabited the former imperial lands granted to them. Unlike the Avars, who apparently maintained a segregation between themselves and subservient non-Avar subjects, the Bulgars integrated the Slavic tribal leadership into their ranks (in a secondary position), opening the door to ethnic intermingling. The resulting Bulgar state was the first officially recognized "barbarian" state in the Balkans and quickly became Byzantium's principal rival for Balkan political hegemony.

Moravians
Slovaks
K. of FRANKS
Danube R.
East Slav Tribes
AVAR KAGHANATE
LOMBARD KINGDOM
Slovenes
Pannonian Slavs
Ravenna
Croats
Srem
Belgrade
BULGAR STATE
Split
Serbs
Vidin
Danube R.
Silistra
Niš
Pliska
The Seven Slav Tribes
Varna
Dubrovnik
Nesebŭr
Pomorie
Sofia
LOMBARD KINGDOM
Skopje
BYZANTINE
Plovdiv
Adrianople
Durrës
Ohrid
Naples
Constantinople
Thessaloniki
Vlorë
EXARCHATE of RAVENNA
Kuber's "Bulgars"
EMPIRE
MILES
0 50 100 150 200
Athens
Patras
Corinth
0 100 200 300
KILOMETERS
Border of Byzantine Empire, end of 7th Cent.
Approximate boundary of Kuber's "Bulgar" authority
Territories under direct Byzantine control
Byzantine territories settled by South Slav Tribes

Map 8: Rise of the First Bulgarian Empire, 7th–10th Centuries

The treaty of 681 with Byzantium granted the Bulgars territories south of the Danube and north of the Balkan Mountains. From their capital at Pliska, the Bulgar rulers controlled the Slavic inhabitants of their newly acquired lands, as well as Wallachia and other lands north of the Danube stretching northeastward to the Eurasian steppes. Under the successive heirs of Asparuh, Bulgaria (also known as the First Bulgarian Empire) attempted to expand its territories in the Balkans at Byzantine expense, either through peaceful and favorable alliances with the imperial authorities, such as Tervel's (701-18) with Emperor Justinian II (705-11), or through warfare. By the opening of the 9th century, Bulgar ruler Krum (808-14), from his capital at Pliska, ruled a large state stretching to Great Moravia in the north and was in the position to commence life-threatening attacks on the Byzantine Empire.

After crushing Avar forces in the Danubian Basin (803), Krum was determined to establish absolute autocracy when he emerged as Bulgar ruler in 808. Throughout his reign, Krum conducted all-out warfare against Byzantium for control of the Balkans. Although his military successes at first fluctuated, in his final years he generally prevailed. In 811 the Bulgars massacred a Byzantine army led by Emperor Nicephoros I (802-11), who was killed and whose head was ordered to be fashioned into a silver-lined ceremonial drinking cup by Krum. During the next two years he swept the Byzantines from stretches of the Black Sea coast, decisively defeated another imperial army, and appeared before the land walls of Constantinople, briefly causing a panic in the imperial capital. At his death, Krum bequeathed to Bulgaria expanded territories in the central and eastern Balkans, including the Byzantine city of Sofia.

Krum's immediate successors consolidated Bulgaria's gains through an extended peace treaty with Byzantium before initiating conquests in the central and western Balkans. Prince Boris I (852-89) continued to push his borders westward but met sharp resistance from Germans in the Croatian northwest and from Serbs, many of whom then came under direct Bulgar control. Boris also established Ohrid as an important Bulgar administrative center in Macedonia, and he expanded his control over western Thrace in the south until an outlet was opened on the Aegean Sea. Boris is most significant for having initiated the conversion of his mixed Turkic and Slavic subjects to Orthodox Christianity (865). The conversion resulted in the creation of a written form of the Slavic language, the single most important Slavic cultural development in medieval Europe.

Disciples of Cyril and Methodios, two brothers sent on an unsuccessful Byzantine mission to Great Moravia, found refuge in Bulgaria following their expulsion by the Germans, bringing with them a cumbersome Glagolitic alphabet created by Cyril expressly for writing Slavic. Under Boris's patronage, the refugees created a powerful new, simpler Slavic alphabet for freeing Bulgaria from possible Byzantine ecclesiastical, and consequent political, control—Cyrillic. Named in honor of Cyril, Cyrillic's characters precisely represented Slavic phonetics. Greek liturgical texts in Bulgaria were translated into Slavic using the new letters, and Bulgarian clergy were trained in Cyrillic. The resulting native Bulgarian church organization provided Bulgaria with the cultural component to ensure its independence.

The strength of the large, autocratically centralized Bulgar state, which represented the collective legacy of Krum and Boris, was fully realized by Boris's successor, Simeon I (893-927). During Simeon's reign, Bulgaria achieved its historical apex, encompassing virtually all of the Balkan Peninsula except Croatia, Thessaloniki, Greece, and the Thracian environs of Constantinople. His continued royal patronage of the Moravian refugees' literary and educational activities in the capital at Preslav, as well as his personal literary interests (Simeon had been educated as a hostage/monk in Constantinople prior to ascending the Bulgarian throne), sparked a Bulgarian Slavic cultural boom. His reign is termed Bulgaria's "golden age."

Like Krum, Simeon conducted near-constant warfare with Byzantium. Within a year of his succession he placed a cohesive and experienced military force in the field that wrested control of most of the Balkans from the empire. Bulgarian military power forced Byzantium to recognize the Bulgarian church's autonomy (a Bulgarian patriarchate was established in Ohrid), though technically it still was considered subordinate to the Greek patriarch of Constantinople by the Byzantines. The Bulgarian ruler was recognized as second in power and authority only to the emperor in Constantinople. Four times between 913 and 924 Simeon advanced to the land walls of the imperial capital but was unable to breach them. Simeon proclaimed himself emperor—*tsar,* in Slavic—of the Romans and Bulgarians (924), and his claim was taken seriously. Ultimate success appeared within his grasp, and only the impregnable strength of Constantinople's defenses prevented Simeon from being acknowledged as emperor of Byzantium.

While occupied to the south with operations against the Byzantines, Simeon was compelled to deal with military problems on other fronts. Starting in 917, Byzantium attempted to weaken Bulgaria militarily by inducing Balkan Slavic and steppe nomadic Turkic peoples to attack Bulgaria from the north, thus opening operations against Simeon on two fronts simultaneously. The Magyars, a Turkic people, swept out of the steppes and permanently wrested all Bulgarian territories north of the Danube, as far as Pannonia, from Simeon's authority. Serbs, urged on by the Byzantines, rebelled against the Bulgarians in 918, and in 924 Simeon conquered and ravaged Serbian lands formerly outside of the Bulgarian state. Despite their efforts, the Byzantines never succeeded in containing Bulgarian expansion in the central and southern Balkans while Simeon was alive. He died in 927 an imperially frustrated but powerful regional ruler.

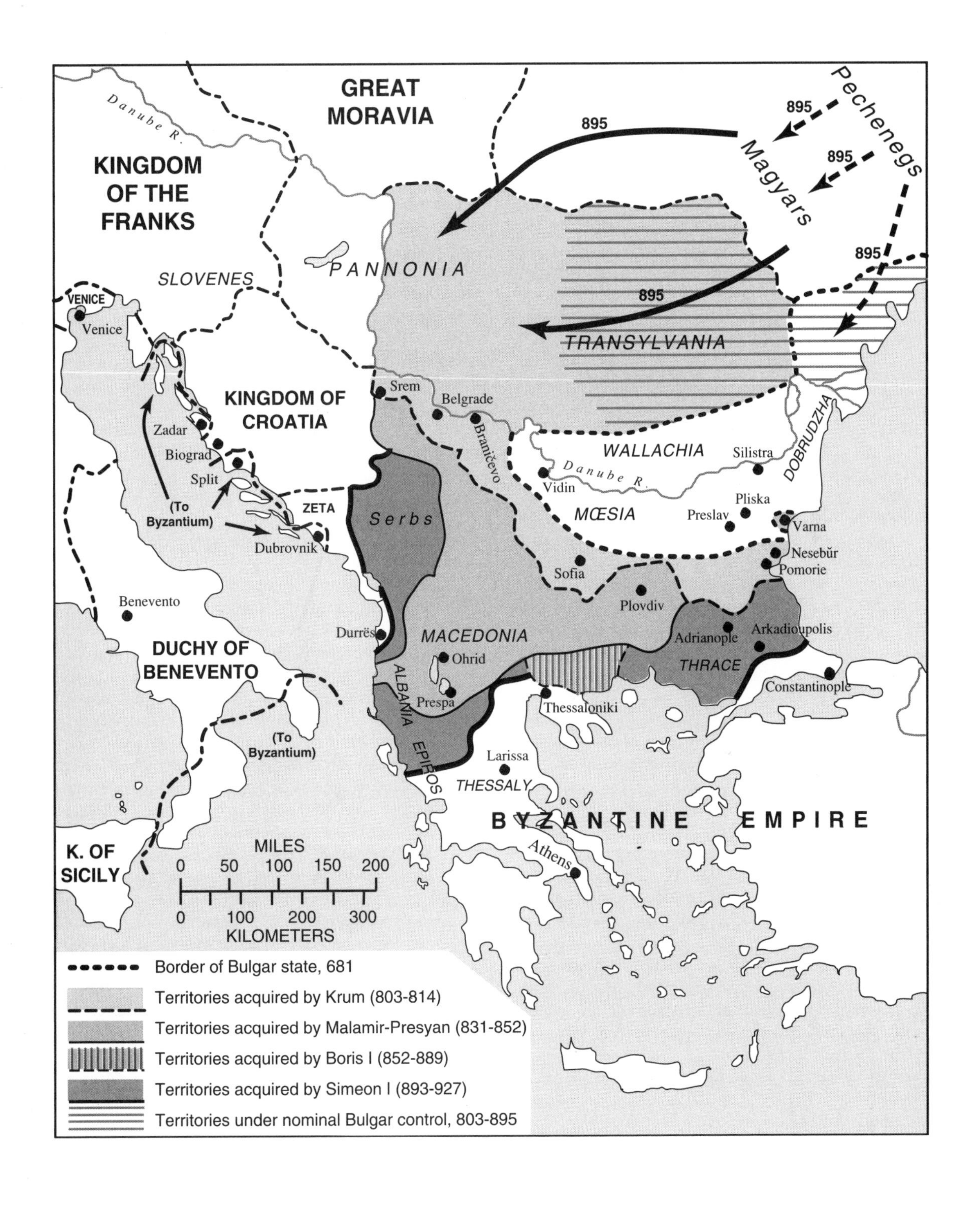
GREAT MORAVIA
KINGDOM OF THE FRANKS
Danube R.
895
Pechenegs
Magyars
PANNONIA
SLOVENES
VENICE
Venice
TRANSYLVANIA
KINGDOM OF CROATIA
Srem
Belgrade
Braničevo
Zadar
Biograd
Split
WALLACHIA
Silistra
DOBRUDZHA
Vidin
Danube R.
MŒSIA
Pliska
Preslav
Varna
Nesebŭr
Pomorie
(To Byzantium)
ZETA
Dubrovnik
Serbs
Sofia
Plovdiv
Benevento
Durrës
MACEDONIA
Adrianople
Arkadioupolis
DUCHY OF BENEVENTO
Ohrid
THRACE
ALBANIA
Prespa
Thessaloniki
Constantinople
(To Byzantium)
EPIROS
Larissa
THESSALY
BYZANTINE EMPIRE
Athens
K. OF SICILY
MILES
0 50 100 150 200
0 100 200 300
KILOMETERS
Border of Bulgar state, 681
Territories acquired by Krum (803-814)
Territories acquired by Malamir-Presyan (831-852)
Territories acquired by Boris I (852-889)
Territories acquired by Simeon I (893-927)
Territories under nominal Bulgar control, 803-895

Map 9: Fall of the First Bulgarian Empire, Mid-10th–Early 11th Centuries

Immediately after gaining the throne, Simeon's successor Tsar Petŭr I (927-67) cemented Bulgaria's claim to imperial standing and the autochephalous status of the Bulgarian Ohrid Patriarchate in a treaty with Byzantium. His marriage to Byzantine princess Maria Lekapena increased Byzantine political and cultural influences in Bulgaria. Little is known about Petŭr's reign until its final years. It seems that ethnic Bulgar numbers and pagan religious practices declined to insignificance, Christian Slavic speakers dominated society, peasants constituted the majority population, and the *tsar* ruled through both edicts and Byzantine-like law codes. Orthodox Christian monasticism flourished during the 930s, producing Bulgaria's most venerated and popular native saint, Ivan Rilski. Petŭr's reign also witnessed the rise of the dualistic Christian Bogomil heresy, which, though often depicted as a mass peasant movement, probably was confined to a literate elite and proved little more than a nuisance to Bulgarian authorities.

Petŭr's good relations with Byzantium did not prevent Byzantine efforts to bring down Bulgaria. Emperor Nikephoros II Phokas (963-69) forged an anti-Bulgarian alliance (965) with the rising state of Kievan Russia, calling on its ruler, Svyatoslav (962-72), to harry the Bulgarians from the north. In 967 Svyatoslav crushed the Bulgarians and occupied much of Bulgaria, forcing Petŭr to abdicate. Nikephoros, concerned over Svyatoslav's decision to remain in Bulgaria, installed Petŭr's son, Boris II (967-71), as a Byzantine client ruler, but Svyatoslav captured Preslav, reduced Boris to subservience, and incorporated Bulgarian forces into his army. Nikephoros's successor, John I Tzimiskes (969-76), attacked Svyatoslav in Bulgaria, took Preslav, captured Tsar Boris, and forced Svyatoslav out (971).

Tzimiskes's triumph was a disaster for Bulgaria. Boris was divested of his imperial position, the Bulgarian Ohrid Patriarchate was abolished, and the state was dismembered. Byzantium annexed Mœsia and Dobrudzha, while the Hungarians took all of Bulgaria's Pannonian and Transylvanian holdings, leaving Wallachia an administrative no-man's land. Only Bulgaria's western regions remained in Bulgarian hands.

Beyond manning the Bulgarian forts along the lower Danube and installing two military governors in Mœsia and Dobrudzha, there was little direct Byzantine presence in annexed eastern Bulgaria, and in Macedonia the Bulgarians virtually were independent. Bulgarian leadership in Macedonia initially coalesced around a certain Count Nicholas's four sons—David, Moisei, Aron, and Samuil—who waited for Tzimiskes to die (976) before moving to unite all of the western Bulgarian lands under their collective authority. By 988, however, Samuil (976-1014) alone ruled over western Bulgaria. Although modern Macedonian nationalists claim Samuil's state as their own, all extant evidence suggests that Samuil considered his state Bulgarian.

Taking advantage of a civil war over the succession to the Byzantine throne, Samuil expanded his control over all western Bulgarian lands, and by 986 he held over half of the eastern Bulgarian lands lost in 971, including the city of Sofia. Byzantine emperor Basil II (976-1025), sitting unsteadily on the throne following the civil war, could ill afford to permit Samuil's successes to go unpunished. He invaded Bulgaria in 986 but failed to capture Sofia, his objective. During his retreat, Basil was ambushed in the mountains and his forces were slaughtered. Any hope for immediate revenge on Bulgaria for the debacle was shelved because Basil lacked troops and faced a rebellion in Anatolia.

Meanwhile, Samuil expanded his holdings. By 997 he controlled all of Macedonia (except Thessaloniki), Mœsia, Epiros, Thessaly, western Thrace, Durrës, and much of present-day Albania. He then turned on the small Serbian states of Zeta (present-day Montenegro) and Raška, which were transformed into client vassals. Following those successes, Samuil had himself crowned *tsar* (997), ruling an enormous empire stretching from the Adriatic to the Black seas and from the Aegean Sea to the Danube River from his capitals at Ohrid and Prespa.

At the turn of the millennium, Basil at last could undertake a concerted effort to crush Samuil. Determined to obliterate Bulgaria, Basil unleashed two decades of relentless, sustained, year-round military offensives. Samuil lost half of his territories, and his troops proved unequal to the Byzantines in stand-up combat. By the end of the first decade of warfare, his Macedonian heartland was caught in a pincers between Basil's gains along the Danube and Byzantine forces in Thessaly. Finally, in 1014 Samuil was forced to make a stand with his main army in a fortified mountain pass east of Petrich. Basil attacked, outflanked the Bulgarian position, surrounded, and crushed Samuil's forces. Although Samuil escaped, some 14,000 of his men were said to have been captured. Basil is said to have had them blinded, except for one out of every hundred men, who was spared one eye to lead the rest back to Samuil's camp. Samuil, stricken by the sight, suffered a massive stroke and died. The victory and its grisly aftermath won Basil the sobriquet of *Boulgaroktonos*—"The Bulgar-Killer."

Recent scholarship dismisses the blinding story as a fiction concocted centuries after the fact. The battle did not prove decisive—the loss of 14,000 men should have been catastrophic for the Bulgarians—and the war dragged on for another four years. Samuil's death, from whatever cause, did bring on a leadership crisis in Bulgaria, resulting in internal discord between successors possessing only localized power bases. When Ivan Vladislav (1015-18) was killed besieging Durrës (1018), Bulgarian resistance finally crumbled. All Bulgarian territories were annexed by Byzantium and divided into three provinces—Bulgaria (Macedonia), Sirmium (the client Serb lands), and Paristrion (Mœsia and Dobrudzha). Byzantine garrisons were stationed in key cities and the annexed Bulgarian territories were firmly fixed in Byzantine hands.

POLAND
KIEVAN RUSSIA
Danube R.
Vienna
HOLY ROMAN EMPIRE
Esztergom
Oğuzes
HUNGARY
Kalocsa
Zagreb
CROATIAN PANNONIA
TRANSYLVANIA
Venice
Pechenegs
To Venice
CROATIA
Belgrade
Biograd
BOSNIA
Vlahs
Split
Danube R.
Silistra
SERBIA
HOLY ROMAN EMPIRE
ZETA
BULGARIA
MŒSIA
Preslav
Dubrovnik
Duklja
Sofia
D. of BENEVENTO
MACEDONIA
Plovdiv
Bari
Petrich
Naples
Durrës
Ohrid
Adrianople
Constantinople
Saracens
Thessaloniki
THRACE
Prespa
THESSALY
Larissa
BYZANTINE EMPIRE
MILES
0 50 100 150 200
K. of SICILY
Athens
0 100 200 300
KILOMETERS
Corinth
Border of Bulgaria under Tsar Samuil (976-1014)
Bulgarian lands conquered by Byzantium before 976
Byzantine Empire after 1018
CRETE

Map 10: The Rise of Medieval Croatia, 10th–12th Centuries

During the 790s forces of Frank king Charlemagne (771-814) attacked and defeated the Avar *kaghanate*, winning Charlemagne control over western Pannonia, northern Dalmatia, and Slavonia, which mainly were inhabited by Croats. During the 9th century two fluid Croat states in western Pannonia and northern Dalmatia arose, whose native rulers were recognized as tributaries by the Franks. Although technically under Frank suzerainty, Dalmatian Croatia drew closer to the Byzantine orbit, since Byzantium had reestablished its direct control of most Dalmatian coastal cities by that time. An active Byzantine presence in Dalmatia declined, however, as a result of Byzantium's struggles with Simeon's Bulgaria, and Dalmatia essentially was left to its own devices.

A Dalmatian Croat rebellion in 875 ended Frank suzerainty and an independent state was created, though the Franks retained control of Pannonian Croatia for a while longer. The princely throne of the new state proved unstable, with occupants following one another in rapid succession. One successful claimant, Branimir (879-92), pledged his state's loyalty to the pope, from which time Dalmatian Croatia received papal recognition of its independence in return for Rome's spiritual jurisdiction within the state.

Tomislav (ca. 910-28), an ally of Byzantium, gained the throne of Dalmatian Croatia sometime around 910, consolidated an army and navy, and liberated Pannonian Croatia from depredations inflicted by the Magyars (Hungarians), who settled on the Pannonian Plain after 895 and were terrorizing their neighbors. Tomislav incorporated Pannonian Croatia under his rule and established a border with the Magyars along the Drava River, creating the first united Croatian state, which probably included Dalmatia, what is today Croatia Proper, western Slavonia, and the greater part of Bosnia. There was no permanent capital, but Tomislav's chief residence appears to have been Biograd. Around 923 Tomislav concluded an alliance with Byzantium against Simeon. Fearful of being caught between the Byzantines in the south and Croatia in the northwest, Simeon invaded Tomislav's lands in 926 but suffered a resounding defeat. Tomislav emerged as the ruler of a militarily strong state. With his death in 928, however, united Croatia disintegrated in civil war.

In his search for anti-Bulgarian allies during his wars with Samuil, Basil II turned to the ruler of Croatia, Stjepan Držislav (969-97), in an attempt to reassert an active Byzantine presence in Dalmatia. Although he was courted by Basil, Držislav proved ineffective as a counter-force against the Bulgarians, and, after he died, Croatia was weakened by a power struggle among his sons. Byzantium, occupied with Samuil in the south, could do nothing to stabilize the situation, and therefore left it to the Venetians to intervene in the region.

Most of the self-governing Dalmatian cities submitted loyalty oaths to Venice, but they also viewed Venice as a commercial competitor whose enhanced position threatened their trade interests. Dubrovnik refused even nominal Venetian control, strengthened its contacts with Byzantium, and won recognition as an imperial province in its own right. With Croatia in disarray, Venice insinuated itself in Croatian affairs. A game of competing nominal suzerainty over Dalmatian coastal cities ensued between Venice and Croatia, dying down in 1019 when Basil II reclaimed all of imperial Dalmatia. Croatia became a Byzantine vassal and Venice fell into civil war.

Kresimir III (1000-30) reasserted control over the Croatian Dalmatian cities and, after Basil's death, ceased paying homage to Constantinople, but the Magyars took Slavonia. Kresimir's successor, Stjepan I (1030-58), preserved Croatian control over its Dalmatian coastal cities and increased Croatian territory. Needing the Croats to defend Dalmatia from a rising Norman threat, Byzantium restored good relations with Croatia during the reign of Petr Kresimir IV (1058-75), who was named commander of Byzantine Dalmatian forces (1069). Kresimir loosely ruled over three provinces—Bosnia, the Slavic Dalmatian coast, and Slavonia. The latter was highly autonomous and governed by a son-in-law of Hungarian King Béla I (1061-63), Zvonimir Trpimirović. Zvonimir accepted annexation by Croatia in return for his continued autonomous rule, an important voice in state matters, and his own succession to the throne should Kresimir die childless. In 1075 Zvonimir (1075-90) was crowned Croatian king following Kresimir's death without a direct male heir.

Zvonimir attempted to disempower the regional Croat nobility by removing local notables from government positions and replacing them with trusted supporters, earning him the nobles' undying opposition. When he died heirless, his Hungarian widow, sister of Hungarian king László I (1077-95), failed to win their support, so László intervened in Croatia to protect her interests, occupying much Croatian territory. László's successor, Kálmán I (1095-1114), continued the attacks, forcing the Croat notables to accept a controversial written deal in 1102, commonly called the *Pacta Conventa,* in which they agreed to accept a dynastic union with Hungary.

According to the agreement, the Croatian notables recognized the Hungarian monarch's sovereignty in exchange for retaining most of their political, legal, and social autonomy within Croatia. Recent scholarship has demonstrated that the *Pacta Conventa* was a 14th-century forgery. What exactly transpired in 1102 remains unknown. In any event, after 1102 the history and fate of Croatia became linked to those of Hungary and Central Europe until the early 20th century.

HOLY ROMAN EMPIRE
HUNGARY
PANNONIAN CROATIA
DALMATIAN CROATIA
SLAVONIA
BOSNIA
BYZANTINE EMPIRE
Serbs
RAŠKA
HUM
ZETA
Drava R.
Sava R.
Danube R.
Tisza R.
Kupa R.
Una R.
Vrbas R.
Bosna R.
Drina R.
Neretva R.
Ibar R.
Drin R.
Pecs
Zagreb
Krk Is.
Nin
Zadar
Biograd
Bribar
Trogir
Split
Belgrade
Raška
Dubrovnik
Kotor
Duklja
Shkodër
Bar
Miles
0 10 20 30 40
0 20 40 60
Kilometers
Croatian Kingdom from Tomislav (ca. 910-928) to Petr Kresimir (1058-1075)
Croatian lands lost to Zeta after 1081
Territories of the Byzantine Empire

Map 11: The Balkans, Late 12th Century

In 1054, the formerly unified European Christian church was torn into two halves—western Roman Catholic and eastern Orthodox—in the Great Schism. The break was the culmination of centuries of rivalry between the Roman papacy and the patriarchate of Constantinople for spiritual supremacy within the church, as well as between the Byzantine and Holy Roman emperors for secular hegemony over a purely theoretical universal Christian world state. Although technically a religious matter, the schism sealed a cultural division between East and West expressed in mutual political animosity and ethnoreligious bigotry, the consequences of which have persisted into the present.

The powerful Macedonian dynasty of the Byzantine Empire had slipped into rapid decline after Basil II, with the provincial military aristocracy pitted against the bureaucratic functionaries of the capital for dominant influence over Basil's successively weak successors. Civil wars broke out in the mid-11th century, and the imperial throne alternated between champions of the provincial military, on the one hand, and functionary factions, on the other, until the able military commander Alexios I Komnenos (1081-1118) led a successful revolt, securing the throne and establishing a stable dynasty (1081-1185).

By the time Alexios returned stability to the imperial office, the empire had suffered frightful military blows at the hands of the Normans in Italy (the last Byzantine foothold was lost in 1071) and the Seljuk Turks in Anatolia (the Byzantine army was irreparably crushed at the Battle of Manzikert in 1071 and much of the region lost). Alexios initiated much-needed judicial and fiscal reforms, won the support of provincial military commanders, and counterbalanced their growing power by playing the high Orthodox clergy against them. He foiled the Italian Normans' attempted invasion of the Balkans (1085), suppressed the Bogomils in the former Bulgarian lands (1086-91), and blunted Cuman incursions south of the Danube (1091). Realizing his military inability to evict the Seljuks from Anatolia, Alexios requested western allies from Pope Urban II (1088-99), thus sparking the Crusades. (See Map 12.)

Alexios received far more than he had bargained for in the Crusades. A succession of western crusading armies marched through the Balkans on their way to Constantinople, from which place they intended to sweep south through Anatolia and liberate the Holy Land from its Muslim Turkish masters. The undisciplined crusaders moved like a horde of human locusts through the Serbian and Bulgarian lands of the empire, cutting swaths of destruction and death as they traveled. Though Alexios managed to shuttle each crusader force into Asia with the least possible delay, he and his successors were unable to prevent or to repair the internal disruption caused by them in the empire's Balkan possessions. Coupled with imperial preoccupation with the crusaders' activities in Asia, that disruption resulted in the reemergence of an independent Bulgarian state in 1185 and the solidification of Serbia under a new royal dynasty by 1196. (See Map 12.)

In 1185 two brothers of possible Vlah origin, Petŭr and Ivan Asen, raised a revolt against the empire in its Bulgarian lands. Though initially defeated and forced north of the Danube, they returned in 1186 with a Cuman army and compelled Emperor Isaac II Angelos (1185-95) to sign a truce giving them control of lands between the Danube and the Balkan Mountains. In 1189, the Asens raided deep into the empire, provoking Angelos to invade their lands in response. The Bulgarians soundly defeated Angelos and forced him to accept the independence of the resurrected Bulgarian state. Both brothers eventually were murdered by members of their unruly aristocracy (Ivan in 1196, Petŭr in 1197), and their younger brother, Kaloyan (1197-1207), succeeded them. He signed a peace treaty with Byzantium (1201) and launched successful campaigns against Serbia and Hungary, extending Bulgaria's western borders.

In Serbia at this time, Stefan Nemanja (ca. 1167-96), a *župan* (local ruler) from the Raška region, managed to unite various clans into a functioning state, which he held together through Orthodox Christianity. Bogomil heretics, who had entered his lands from Bulgaria, were persecuted and forced into Bosnia. Stefan proclaimed his complete independence from Byzantium and expanded his territories to the west and south, incorporating the older Serbian state of Zeta. He abdicated in 1196 and retired to Hilandar Monastery on Mount Athos, which one of his sons, St. Sava (Rastko), had founded. (See Map 17.)

Bosnia had long been a bone of contention between Croats and Serbs, with control passing back and forth between them, until an independent Bosnian state emerged in the late 12th century under the rule of Kulin (ca. 1180-1204). He acknowledged nominal Hungarian suzerainty—while working to attain full independence by commercially developing Bosnia's silver mining industry and establishing lucrative trade arrangements with Dubrovnik—and also parried the powerful cultural-political influences of Roman Catholicism and Orthodoxy by tolerating Bogomilism.

Catholic Hungarian and Orthodox Serbian rulers, eager to gain Bosnia for themselves, accused Kulin and his family of being devoted to the Bogomil belief. They asserted to the pope that Kulin was attempting to make Bogomilism the state religion of the Bosnian lands under his authority, lands that were considered Roman Catholic because of their past Hungarian-Croatian connection. When Pope Innocent III (1198-1216) preached a crusade against Kulin and the heretics, the Bosnian ruler felt constrained to announce his adherence to Catholicism and to permit a Catholic synod in Bosnia condemning Bogomilism (1203). (See Map 16.)

Danube R.
KIEVAN
RUSSIA
HOLY
ROMAN
EMPIRE
Vienna
Esztergom
HUNGARY
Kalocsa
Zagreb
SLAVONIA
TRANSYLVANIA
Cumans
Venice
To
Venice
CROATIA
Belgrade
Zadar
BOSNIA
Vlahs and Cumans
PAPAL
STATES
Split
SERBIA
Vidin
Danube R.
Raška
Tŭrnovo
Dubrovnik
BULGARIA
Varna
(To
Byzantium)
Duklja
Sofia
KINGDOM OF THE
TWO SICILIES
MACEDONIA
Plovdiv
Adrianople
Naples
Durrës
Ohrid
Serres
Constantinople
Thessaloniki
Ioannina
BYZANTINE EMPIRE
MILES
0 50 100 150 200
SICILY
Athens
0 100 200 300
KILOMETERS
CRETE
Border of Second Bulgarian Empire, 1197
Border of resurrected Bulgarian state, 1187
Territories under direct Bulgarian control
Territories under loose Bulgarian control

Map 12: Crusades in the Balkans, Late 11th–Early 13th Centuries

Norman, Serbian, Pecheneg, Cuman, and Seljuk Turk threats during the second half of the 11th century placed the Byzantine Empire in dire straits. In desperation, Emperor Alexios I wrote to the pope in 1090 requesting Western mercenary troops to help bolster his faltering military. Pope Urban II used his letter to strengthen the papacy's temporal authority in the West by placing it at the head of a Christian crusading movement against the East. Alexios received more than he bargained for as a result.

Scholars debate whether Alexios's request for military aid was necessary. By the time the First Crusade arrived in the Byzantine Balkans in 1096, Alexios's military, administrative, and ecclesiastical reforms had stabilized the Byzantine position, and the empire was poised to win back much of its losses unassisted. Criticism from hindsight is easy. In 1090 the empire had been reduced mainly to its Balkan territories. Serbs, Croats, and Venetians steadily stripped away portions of the northwestern Balkans, and chaos caused by Pecheneg and Cuman incursions made Byzantium's hold on the rest shaky. Alexios cannot be faulted for believing that he needed serious military assistance.

Though the Great Schism between Orthodox and Roman Catholic Christianity occurred in 1054, it initially was an ecclesiastical division only marginally affecting the general populations of East and West. That situation changed with the Crusades. Before the first Western feudal forces arrived in the Balkans on their eastward overland trek, there appeared an undisciplined horde of peasants, beggars, ruffians, and soldiers, led by Peter the Hermit, who swept through the Balkans along the Diagonal Highway to Constantinople like a plague of locusts. Alexios quickly ferried them across the Bosphorus to Anatolia and swift extermination by the Seljuks. The ill will against Catholics that they spawned among the Orthodox inhabitants along their line of march became a lasting legacy.

Anti-Catholic sentiment among the Orthodox inhabitants was reinforced when the French and German forces of Duke Godfrey de Bouillon and his brother Baldwin, one of four main contingents comprising the First Crusade, followed the same Diagonal route on their way to join up at Constantinople with the other forces, who used the *Via Ignatia* route running from Durrës. Alexios expected all of the crusaders to use the *Via Ignatia* route, so supplies were not prepared along the Diagonal. Acting more like they were operating in enemy country than in that of Christian allies, Godfrey's troops looted their way south, arriving at the Byzantine capital only after fighting several skirmishes with local Serbian and Bulgarian inhabitants, as well as with Byzantine troops sent to guard the road and maintain order. Although problems between crusaders and natives were fewer on the *Via Ignatia* route, the First Crusade's Balkan marches engendered a rising level of popular animosity between Orthodox and Catholics, making any future compromise on religious differences problematic.

Alexios never wanted the 50,000-strong crusader army that answered his appeal for mercenary troops. Unlike mercenaries, who fought under their employer's orders, the crusaders considered themselves independent of the emperor. Rather than reclaiming Anatolia for Byzantium, they sought only fighting, fame, and fortune in the Holy Land. Alexios and the crusade's leaders suspected one another's motives and intentions. This held true especially for the Normans, Byzantium's inveterate enemies. Even though Alexios obtained Western-style feudal oaths from the crusaders' leaders and assisted them during their Anatolian operations against the Seljuks, Byzantine-crusader relations were strained almost from the beginning and would remain so throughout the crusading epoch, despite sporadic periods of cooperation. The march through the Balkans of the Western forces during the Second Crusade (1147-49) was accompanied by rising levels of conflict with the native inhabitants, and the Third Crusade's troops of Holy Roman emperor Frederick I Barbarossa (1152-90) virtually fought their way along the Diagonal Highway on their way to Anatolia. The growing mutual distrust and antagonism that emerged among the crusaders' common ranks and Byzantium's general population ultimately intensified into bitter cultural animosity between European Orthodox and Catholic Christians.

Orthodox Byzantines, Serbs, and Bulgarians had experienced Catholic crusader looting and violence in locales through which those warriors had marched. The people had been forced to tolerate crusader disrespect for their traditions and insults to their faith, and they came to view the Latin crusaders as a papal threat to Orthodoxy itself. Suspicion of the Orthodox Easterners also grew in the West. Although the Crusades needed cooperation between both branches of Christendom to succeed, some Westerners condemned the Byzantines as schismatics and considered Constantinople a legitimate target for a crusade, believing that fighting "heretics" was the spiritual equivalent of combatting Muslim "unbelievers." Therefore, assaulting Byzantium could be justified religiously as a "crusade."

Pope Innocent III was determined to reunite the two branches of Christendom on Catholic terms and to launch a fourth crusade to recover the Holy Lands for Christianity. In 1199 he called for a new crusade, but his appeal was met enthusiastically only among the nobility of northern France, who contracted with Venice for ships to transport their anticipated crusading army to the East. A series of fateful circumstances, combined with the crusaders' anti-Orthodox/anti-Byzantine sentiments, proved decisive in changing the crusade from an attack on the Muslims in Egypt to an assault on the Byzantine Empire's capital of Constantinople. (See Map 13.)

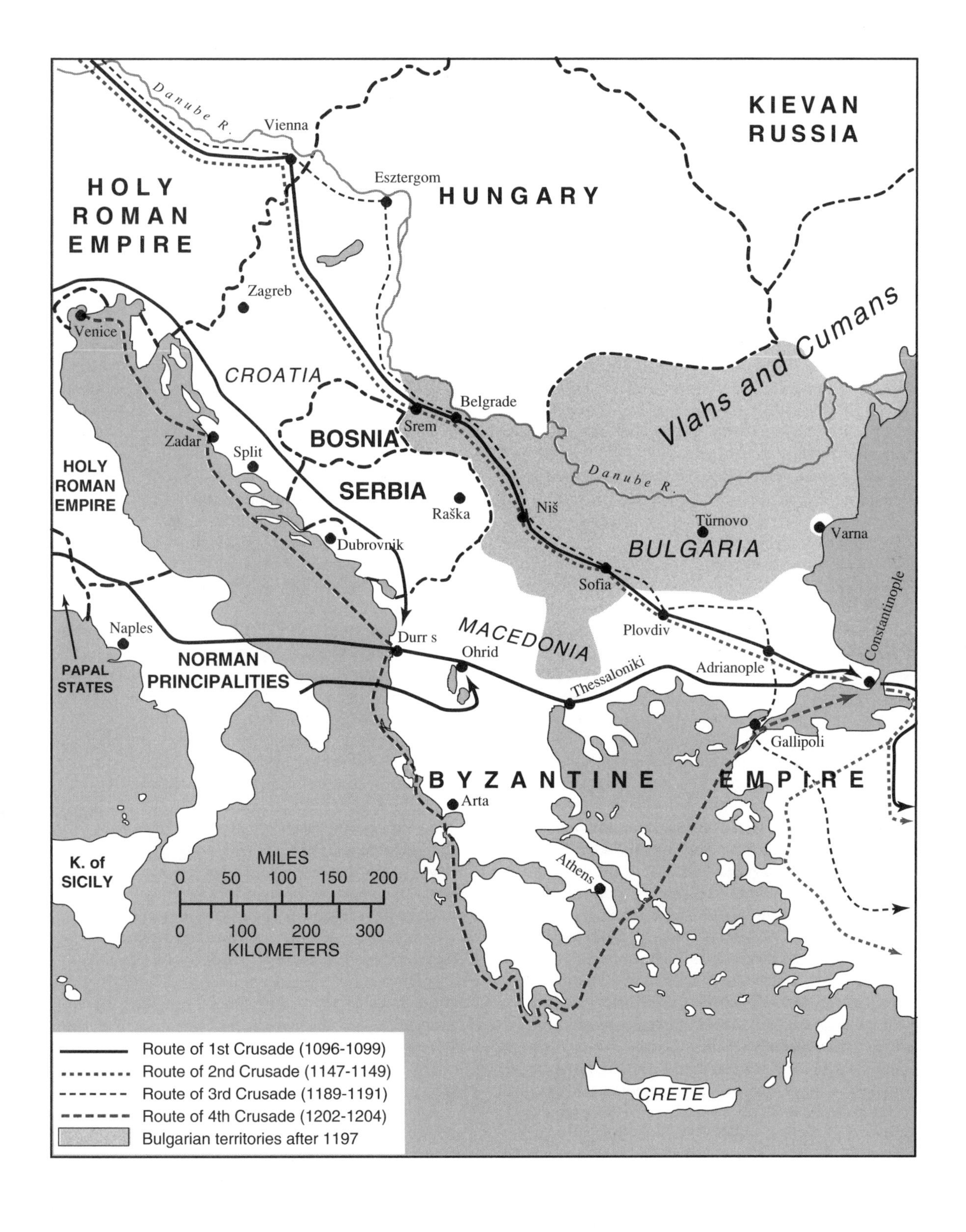
KIEVAN RUSSIA
HOLY ROMAN EMPIRE
HUNGARY
Danube R.
Vienna
Esztergom
Zagreb
Venice
CROATIA
BOSNIA
Vlahs and Cumans
Belgrade
Srem
Zadar
Split
HOLY ROMAN EMPIRE
SERBIA
Raška
Niš
Danube R.
Tŭrnovo
Varna
Dubrovnik
BULGARIA
Sofia
Constantinople
Naples
Durr s
MACEDONIA
Plovdiv
Ohrid
PAPAL STATES
NORMAN PRINCIPALITIES
Thessaloniki
Adrianople
Gallipoli
BYZANTINE EMPIRE
Arta
K. of SICILY
MILES
0 50 100 150 200
0 100 200 300
KILOMETERS
Athens
CRETE
Route of 1st Crusade (1096-1099)
Route of 2nd Crusade (1147-1149)
Route of 3rd Crusade (1189-1191)
Route of 4th Crusade (1202-1204)
Bulgarian territories after 1197

Map 13: The Balkans after the Fourth Crusade, 1204-1214

Catastrophe struck the Byzantine Empire in 1204. Venetian mariners, commanded by Doge Enrico Dandolo (1193-1205), the central figure and mainspring of the undertaking, and French knights, led by Boniface di Montferrat, descended on Constantinople in 1203. They constituted most of the Fourth Crusade's military forces, originally despatched by Pope Innocent III against the Muslim Turkish and Arab masters of Egypt and, later, the Holy Lands. Normans from the Kingdom of the Two Sicilies augmented their ranks. From pope to common warrior, the crusade was riddled with enemies of the Orthodox Byzantine Empire. (See Map 12.)

The cultural animosities engendered in the Great Schism, combined with fateful circumstances, caused the crusade to end with the capture and sack of Constantinople rather than the assault of Muslim Egypt or Palestine. The crusading army that gathered at Venice (1202) in response to Pope Innocent's call proved smaller than expected, but its leaders had contracted ships in advance based on their original estimates. Dandolo convinced Montferrat that the smaller forces could defray Venice's costs by helping him capture the Hungarian-Croatian Adriatic port of Zadar. Once this was accomplished, Dandolo managed to divert the forces to Constantinople. He wanted Venice to win greater commercial advantage in the eastern Mediterranean over Byzantium, to which the city had long been subordinated. Alexios Angelos, son of recently deposed Emperor Isaac II Angelos, aided Dandolo in his scheme by appearing outside Zadar bearing letters from a contender for the Holy Roman imperial throne—Philip of Swabia, son-in-law of the deposed Isaac and brother of late Holy Roman emperor, Henry VI (1190-97)—requesting that the crusaders help restore Isaac to the Byzantine throne.

Under pretext of restoring Isaac, in return for his financial and material support of their crusading efforts, the crusaders descended on Constantinople in 1203. Although they achieved their short-term goal, their champion proved too much a creature of the Orthodox East to play the role of lackey for Catholic Westerners, whom the Byzantines looked upon (with some justification) as barbaric and less culturally developed. In 1204 the Western crusaders turned their greed and frustration on the Byzantines, whom they considered (just as justifiably) to have reneged on promises of support for their endeavor. The crusaders eventually broke through the sea walls protecting Constantinople and, once inside, gave free rein to a venomous cultural animosity that has been rivaled in Southeastern Europe only by the carnage displayed in the recent warfare in Bosnia and Kosovo. The wholesale pillage inflicted by the Catholic warriors on the stricken capital of the East, the largest and wealthiest city in the world at the time, was unprecedented. The Orthodox world never forgot or forgave the West for the sack of Constantinople, and the event created a gulf between the Eastern and Western European civilizations that has existed virtually unbridged into the present.

The Latin Empire founded by the crusaders was able to plant few roots in the hostile Orthodox East and survived for less than 60 years. In feudal fashion, the victors divided the conquered state among themselves. Baldwin of Flanders was proclaimed emperor of Constantinople (1204-5) and a Venetian, Pier Morosini, was raised to Latin (Catholic) patriarch of the East. Montferrat was granted the Kingdom of Thessaloniki, and the rest of the territorial spoils were distributed as the new emperor's vassal holdings to various warriors. Dandolo and Venice received control of a part of Constantinople; most of the Aegean islands; a portion of Achaia; Crete; and possessions in the Adriatic (including Dubrovnik, Durrës, and Zadar). For as long as it existed, the Latin Empire was progressively weakened by feudal rivalries among its constituent territorial lords and vassals.

Three Orthodox states quickly emerged as contenders for reestablishing the stricken Byzantine Empire. The first was the so-called Empire of Nicæa, founded by refugees from Constantinople and headed by the ejected Emperor Alexios V Doukas (1204), that was established in Anatolia. Under their emperor, Theodore I Laskaris (1204-22), the Nicæans effectively prevented the Latins from gaining any permanent foothold in that Asian region.

In the Balkans, newly restored Bulgaria emerged as a leading contender for restoring an Orthodox empire in the East. Its ruler, Kaloyan, defeated Latin Emperor Baldwin and his forces at Adrianople in 1205, capturing the vanquished ruler and holding him in captivity until his death at Tŭrnovo. Bulgaria then expanded south and west at the expense of the Latin Empire. Kaloyan toyed with Catholicism, winning papal recognition of an autonomous Bulgarian church and of himself as king. In 1218 his son, Ivan II Asen (1218-41), overthrew a usurper of the Bulgarian throne and initiated the apogee of the second Bulgarian state. Though personally mild and generous, Ivan proved an effective military commander and statesman. He further undermined the Latin presence in the Balkans, forged successful alliances with Nicæa, definitively broke with Rome (1232), won recognition of an independent Bulgarian Patriarchate of Tŭrnovo from the Greek church, and crippled the third major contender for a Byzantine restoration, the Despotate of Epiros.

Epiros was founded by Michael Angelos Komnenos (1204-14) and expanded along the western Balkan Adriatic coastline at Latin, Venetian, and Bulgarian expense under Theodore Doukas Angelos (1214-30), who ended the Kingdom of Thessaloniki in 1222, proclaimed himself emperor, and defeated a Nicæan force near Adrianople (1224). Defeat and capture by the Bulgarians snuffed out his successes, and the influence of Epiros in the struggle for Byzantine restoration was reduced thereafter. (See Map 14.)

HOLY ROMAN EMPIRE
Danube R.
Vienna
Esztergom
KIEVAN RUSSIA
HUNGARY
TRANSYLVANIA
Zagreb
Venice
SLAVONIA
CROATIA
To Venice
BOSNIA
Belgrade
PAPAL STATES
Zadar
BULGARIA
Raška
Niš
Danube R.
SERBIA
Dubrovnik
Tŭrnovo
KINGDOM OF THE TWO SICILIES
Skopje
Sofia
To Venice
Naples
Durrës
MACEDONIA
Adrianople
Ohrid
THRACE
Constantinople
EPIROS
K. OF THESSALONIKI
Thessaloniki
Nicæa
Larissa
CORFU
Arta
THESSALY
EMPIRE OF NICÆA
D. OF ATHENS
EUBŒA
SICILY
PR. OF ACHAIA
Athens
Mystras
SELJUK EMPIRE
MILES
0 50 100 150 200
0 100 200 300
KILOMETERS
RHODES
CRETE
The Latin Empire
Venetian island possessions
Holy Roman Empire

Map 14: Byzantium Resurrected, 1261-1328

By the mid-13th century, only Nicæa, of the three contenders for reestablishing an Orthodox *imperium* in the East, enjoyed any hope of success. Bulgaria had fallen into internal political dissolution and vassalage to the Mongol-Tatars following Ivan II Asen's death in 1241, and Epiros's defeat by Bulgaria reduced it to a marginal role in the western Balkans. In July 1261 a Genoese fleet ferried a Nicæan army across the Bosphorus to Thrace. A reconnoitering force found Constantinople virtually undefended—most Latin and Venetian forces were off besieging a Black Sea island. Sympathetic supporters inside of the city opened an undefended portal in the land walls and the Nicæan army entered. The Westerners fled and the Latin Empire came to an anticlimactic end. Nicæan emperor Michael VIII Palaiologos (1259-82) entered Constantinople in triumph and proclaimed the reestablishment of the Byzantine Empire.

Restored Byzantium was a hollow shell of its former self, encompassing a corner of northwestern Anatolia, Thrace, Macedonia, and a smattering of small holdings in the Peloponnese. The remaining pre-Latin imperial lands lay divided among the Despotate of Epiros, a Greek breakaway Duchy of Thessaly (Neopatras), the Latin states of Athens and Achaia, and the Venetian Duchy of the Archipelago, which included a few ports in the Peloponnese and numerous Aegean islands. Michael devoted his entire reign to recovering those lost territories.

In 1262 Achaia turned over to Michael the Peloponnesian fortresses of Mystras and Monemvasia, which provided bases for Byzantine recovery of Morea. Michael established a district of the Morea governed from Mystras and, by the first decades of the 14th century, Byzantine Mystras controlled most of the southern Peloponnese. Much of the Despotate of Epiros was occupied, and the rest of it was forced to recognize Byzantine suzerainty (1262). By 1265, much of northern Thrace once again was in Byzantine hands. Michael's gains were consolidated by marriage alliances contracted with the rulers of Epiros, Bulgaria, and the Golden Horde. Hungary was made an ally to serve as a Damocles sword should Serbia attempt to cause trouble in the northwest.

The greatest threat to Byzantium came from the West, where the papacy and the Anjevins of the Kingdom of the Two Sicilies were eager to reinstate the Latin Empire in Constantinople. Charles I de Anjou (1262-85) made lengthy preparations for the venture by gathering an invasion fleet and creating a web of anti-Byzantine alliances with various Balkan states. Michael stymied the threat by adept diplomacy. He sent a Byzantine delegation to the Catholic church Council of Lyon (1274), where they acknowledged both papal primacy and the union of the Orthodox and Catholic churches. In exchange, Michael obtained papal assurances of a free hand in reconquering all formerly Byzantine territory in the Balkans. Internal opposition within the Byzantine Empire, however, blunted the advantages Michael briefly gained at Lyon. The majority of the empire's Orthodox clergy and population repudiated the union, causing a veritable schism within the Orthodox church, and a new, pro-Anjevin pope condemned Michael as a schismatic and declared him deposed.

With Michael isolated on all sides, Charles attempted to reopen his long-planned Balkan invasion. That opportunity ended when the wily Byzantine emperor pulled off his most successful diplomatic feat. In 1282 Byzantine gold stirred up the Sicilian Vespers revolt against Charles in Sicily, while Michael financed a fleet for King Pedro III (1276-85) of Aragon to assault Sicily and wrest the throne from Charles. Although Charles retained his mainland Italian possessions, he permanently abandoned his invasion plans.

Michael paid dearly for his efforts to reclaim and defend Byzantium's position in the Balkans. He antagonized his own people by the church union; his constant military efforts were conducted at the expense of the empire's Anatolian defenses; and his army's losses left it dependent on Latin and Turk mercenaries. Rapid decline soon followed his death in 1282. Emperor Andronikos II Palaiologos (1282-1328) failed to prevent territorial losses to Serbia in Macedonia and northern Albania. Although diplomatic steps, such as renouncing the Lyon church union, brought Byzantium internal reconciliation and a small measure of external security, the sad state of the military bequeathed him by Michael left Andronikos nearly helpless in the face of rising Serbian expansion. (See Map 17.)

The empire's poor financial situation constrained Andronikos II to cut costs, with particularly devastating effect on the military. Economic dependence on Genoa led to placing the empire's naval defenses completely into those foreigners' hands. Since paid mercenaries constituted the army, Andronikos drastically reduced their numbers. In Anatolia, the Ottoman Turks persistently encroached on imperial lands while, in the Balkans, Byzantium was hard-pressed to withstand periodic Serbian and Bulgarian raids. Creative tax reforms increased state revenues, but these were squandered on bribes intended to keep neighboring rulers in check.

An attempt to use the mercenary Catalan Grand Company from Sicily (1304) against the Ottomans backfired when their pay went into arrears. After their leader Roger de Flor was assassinated by Byzantine agents (1305), the Catalans ran amuck in Thrace. They ravaged Thrace for two years, then moved into Macedonia, where their depredations continued (oftentimes in partnership with Serbia), and, eventually, into Thessaly, which they pillaged for a year before seizing the Duchy of Athens for themselves (1311). With the passing of the Catalans, the last Western threat to Byzantium ended, but the empire was left ravaged and weak. Byzantium no longer was a great power in its own right.

HOLY ROMAN EMPIRE
Danube R.
Vienna
Esztergom
HUNGARY
KIEV
GOLDEN HORDE
BESSARABIA
Suceava
MOLDAVIA
TRANSYLVANIA
VENICE
Venice
ISTRIA
Zagreb
CROATIA
SLAVONIA
Curtea de Argeş
Braničevo
Belgrade
BOSNIA
WALLACHIA
DOBRUDZHA
Danube R.
Zadar
Split
PAPAL STATES
RAŠKA
HUM
SERBIA
Raška
Niš
Tŭrnovo
Varna
Dubrovnik
ZETA
Duklja
Sofia
BULGARIA
KOSOVO
Kyustendil
Plovdiv
KINGDOM OF THE TWO SICILIES
Naples
Durrës
To Sicily
Skopje
Adrianople
Ohrid
MACEDONIA
Serres
THRACE
Constantinople
ALBANIA
Thessaloniki
BYZANTINE EMPIRE
THESSALY
Arta
EPIROS
Larissa
MILES
0 50 100 150 200
0 100 200 300
KILOMETERS
SICILY
D. OF ATHENS
Patras
Athens
PR. OF ACHAIA
Mystras
MOREA
Monemvasia
SELJUK EMPIRE
CRETE
Byzantine Empire
Venetian possessions

Map 15: Rise of the Romanian Principalities, Mid-13th–14th Centuries

Often collectively identified as the Romanian (or Danubian) Principalities, Wallachia and Moldavia emerged as separate political entities. The political independence of both was precarious and preserved only by their status as veritable satellites of their powerful neighbor, Hungary. The shared Orthodox culture of the two, acquired through some of their populations' connections with the Second Bulgarian Empire, remained closely linked to developments south of the Danube.

Vlahs and Cumans living on the Wallachian Plain and the Moldavian tablelands played a role in restoring an independent Bulgaria in the 1180s, but Bulgarian political control over them was nominal. Slavic Orthodox Christianity prevalent in Bulgaria took firm hold among the semipastoral Vlahs in the 13th century, after which they adopted a more settled lifestyle. Since the Vlahs' native language was Latin-based, their acceptance of the Slavic liturgy and the use of the Cyrillic alphabet in writing their vernacular bespoke strong cultural ties with Bulgaria and Orthodox European civilization in general. Any Bulgarian control over them, however, was obliterated by the Mongol-Tatar devastation of Bulgaria in 1241-42.

In the 1240s the Hungarians pushed southward onto the Wallachian Plain to stifle Cuman and Vlah threats to their borders. There they encountered small Vlah political formations under the authority of local leaders *(voievods)*, which were turned into vassal client districts. By the late 13th century the Wallachian Vlah client districts commenced a gradual process of consolidation, which was accelerated by events inside of Hungary. The end of the native Hungarian Árpád ruling dynasty threw the state into political disarray. Unsettled conditions in Hungary permitted Voievod Basarab (ca. 1310-52), possibly a Cuman prince, to unite the Vlah lands lying between the Carpathians and the Danube under his authority. He negotiated alliances with Bulgaria and Serbia and broke his ties of vassalage to the new Hungarian king Charles Robert I de Anjou (1301-42). In 1330 Charles Robert invaded south of the Carpathians to reimpose Hungarian suzerainty over the region, but he was defeated and forced to recognize the independence of Basarab's state, which acquired the name of Wallachia—the "Land of the Vlahs."

From his Carpathian capital at Curtea de Argeş, Basarab's successor, Nicolae Alexandru (1352-64), obtained Byzantine recognition of an autonomous Wallachian Orthodox church in 1359, when his capital's bishop was elevated to the rank of Wallachian metropolitan by the patriarch of Constantinople. The new metropolitan's jurisdiction was not confined to Wallachia alone but extended over all of the Orthodox Vlahs living in Hungarian-controlled Transylvania. Wallachian independence was strengthened under Alexandru's successor, Vladislav I Vlaicu (1364-77), who issued the first Wallachian coins, forged increasing economic ties with Transylvania, and founded monasteries throughout his territories to serve as centers of Orthodox culture.

The political development of the Vlahs inhabiting the Moldavian tablelands and foothills east of the Carpathian Mountains was impeded somewhat, relative to that of Wallachian Vlahs, because of their vulnerability to incursions by Pechenegs, Oğuzes, Cumans, and Mongol-Tatars from the Ukrainian steppe. In addition, Slavic Ruthenians and Poles were pushing gradually southward into the northern territory (later known as Bukovina) of the region. By the late 13th century, however, there existed small political entities of Orthodox Vlahs headed by local notables throughout Moldavia. During the first half of the 14th century Hungary's presence in the region increased, as it attempted to create border marches east of the Carpathians to protect Transylvania against Mongol-Tatar raids. Although some local Vlah district *voievods* resisted Hungarian control by aligning with the Poles in the north, while others in the southeast came under Mongol-Tatar control, by mid-century the Hungarians had established a vassal Vlah border march in western Moldavia.

Soon after the Hungarian border march in Moldavia was founded, some Maramureş Vlah notables, led by Voievod Bogdan I (1359-ca. 1365), staged an anti-Hungarian uprising that spread into Moldavia. Hungarian king Louis I de Anjou (1342-82) sent an army to crush the rebellion in 1365 only to be defeated, resulting in the de facto establishment of Moldavian independence. Under Bogdan's successors during the remaining years of the 14th century, the young rebel state, ruled from its capital at Suceava, consolidated its control over the territories lying between the eastern Carpathian Mountains and the Prut River, as far south as the Danube frontier with Wallachia and as far north as the southern border of Poland.

Moldavian independence was precarious. Constant political and cultural pressures were exerted on the state by neighboring Roman Catholic powers—Hungary to the west and Poland to the north. The 14th-century Moldavian rulers attempted to counter the dual political threat by playing the two against one another through a complex foreign policy of shifting alliances. After Hungarian king Louis I acquired the Polish throne (1370-82), creating a dynastic union of Moldavia's two primary antagonists, the Moldavian rulers began lengthy negotiations with the Byzantine Patriarchate of Constantinople for the creation of an independent Moldavian Orthodox church. In 1401 Voievod Alexandru the Good (1400-32) finally won Byzantine recognition of his capital's bishop as metropolitan of an autonomous church organization, thus acquiring Moldavia's political recognition within the Orthodox European community.

Romanian Principalities

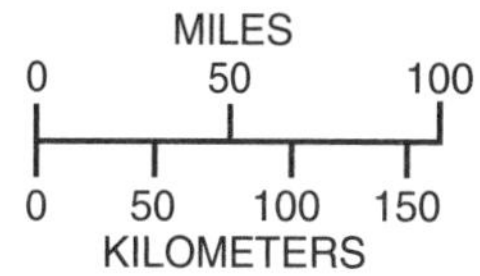

Map 16: Rise of Medieval Bosnia, 13th-14th Centuries

Bosnia was a region lying within the Dinaric Alps in the Balkans' northwest. Originally, it was an inland component of Byzantine Dalmatia but was separated from the Adriatic coastal region during the 11th-century consolidation of Croatia (see Map 10), at which time it acquired the name "Bosnia"—the land centered on the Bosna River. After Croatia's union with Hungary in 1102, Bosnia fell into Hungarian hands during the 1120s. Emperor Manuel I Komnenos (1143-80), however, reestablished Byzantine authority over the region in the 1160s. Following his death in 1180, an independent Bosnian state emerged under a certain Kulin. (See Map 11.)

From Kulin's time on, Bosnia was accused of being riddled with Bogomilism by Catholic outsiders, especially Hungarians, who used the argument to justify repeated but unsuccessful "crusades" in efforts to reassert their control over the region. The Hungarians did succeed in having the pope place all Bosnian Catholics under the authority of their bishop of Kalocsa. In response, the Bosnians rejected the new ecclesiastical administration and established an autonomous Bosnian church of their own, which both the papacy and the Hungarians stigmatized as Bogomil. Existing evidence from Bosnia itself, however, indicates that the Bosnian church was Roman Catholic in dogma, though intentionally schismatic in rejection of papal control.

Ban Stjepan Kotromanić (ca. 1318-53) made Bosnia an important player in the northwestern Balkans' political affairs. Kotromanić gained recognition of Bosnia's continued autonomy under a very loose nominal Hungarian suzerainty. He solidified good relations with the Hungarians by marrying his daughter to Hungarian king Louis I de Anjou. He also successfully resisted a Serbian invasion led by Car Stefan Uroš IV Dušan (1331-55) in 1350 and expanded the territory under his authority to include parts of Croatia and most of Hum (today's Hercegovina), gaining control of the Neretva River valley to the Adriatic.

It was under Kotromanić that a full-fledged autonomous Bosnian church emerged. Headed by a bishop with the title *djed* (grandfather) and staffed by clergy drawn exclusively from Catholic monasteries, the Bosnian church enjoyed close ties with the *ban*. Outside of Bosnia, the papal claim that the Bosnians were heretics persisted. Franciscan missionaries were sent to reassert papal authority in Bosnia. Kotromanić, originally Orthodox, converted to Catholicism and, from that moment on, all Bosnian rulers were Roman Catholic. Despite the missionaries' success with the rulers, many of Bosnia's local nobility and population persisted in their autonomous Bosnian faith or in Orthodoxy.

Kotromanić developed Bosnia's mining industries further, opening new mines and expanding trade contacts with Dubrovnik. Saxon Germans were brought in from Hungary to provide technical expertise, and Dubrovnik colonists handled the mines' administrative and financial operations. Bosnia prospered as a primary European supplier of silver and lead. Towns flourished, craft industries mushroomed, and commerce thrived. Since the foreign colonists who predominated in the expanding and wealthy Bosnian urban population were Catholic, the Franciscan missionaries found ready support in the thriving towns. The native Bosnian church became restricted to the rural regions that mostly were unaffected by the ore mining boom.

Bosnia reached its apex under Ban Tvrtko I (1353-91). The beginning of his reign was inauspicious. Kotromanić had created a territorially large state but one governed through a weak central administration. At his death, Bosnia began to dissolve into numerous local principalities controlled by independent-minded lords. King Louis of Hungary stripped Tvrtko of much territory before confirming him as his hapless and seemingly dependent vassal *ban*. Contrary to Hungarian expectations, Tvrtko gradually consolidated his authority within Bosnia, despite a noble uprising. By the early 1370s he firmly controlled most of the region.

With his authority assured, Tvrtko began meddling in the affairs of the Serbs lying to his southeast. He took from them nearly all of Hum and the Adriatic coastline between the Gulf of Kotor and Dubrovnik. Having acquired a good deal of Serbian territory and playing on his maternal Serbian ancestry (his grandmother was of the Serbian royal Nemanja family), in 1377 Tvrtko claimed the royal crown of both Bosnia and Serbia. Although few Serb nobles outside of Bosnia's borders recognized his claim, all successive Bosnian rulers used the royal title rather than that of *ban*. Following the death of Hungarian king Louis I in 1382, Tvrtko took advantage of the subsequent regency to extend his control over much of Croatia and Dalmatia, adding the names of both to his assumed royal title.

On Tvrtko's death in 1391, his Bosnian kingdom remained intact, though the central power of his successors was weak relative to the strong regional authority of the local nobles, who sported their own subordinate vassals and military forces. Those lords seem to have maintained the state because they deemed it in their best interests to do so—small local principalities would have been easy prey for the neighboring Hungarians and Serbs. The king exerted effective authority only over the central territories of Bosnia, while the local nobility reigned in areas conquered by Kotromanić and Tvrtko. Such regional particularism was abetted by the existence of three separate Christian religions—Catholicism, Orthodoxy, and the Bosnian church. Each tended to dominate in given areas of the kingdom to the relative exclusion of the others, thus preventing any particular one from serving as a vehicle for strong state integration.

HOLY ROMAN EMPIRE
HUNGARY
CROATIA
SLAVONIA
DALMATIA
SERBIA
RAŠKA
HUM
ZETA
Sava R.
Drava R.
Danube R.
Una R.
Vrbas R.
Bosna R.
Drina R.
Neretva R.
Ibar R.
Zagreb
Pecs
Djakovo
Srem
Belgrade
Jajce
Srebrenica
Visoko
Sarajevo
Zadar
Split
Radimlja
Ston
Dubrovnik
Raška
Duklja
Shkodër
MILES
0 10 20 30 40
0 20 40 60
KILOMETERS
Bosnian borders in 1200
Bosnian borders in 1326
Bosnia in 1391
Venetian possessions

Map 17: Rise of Medieval Serbia, 13th–Mid-14th Centuries

In the mid-11th century the Serbs of the Zeta (Montenegrin) region of the Byzantine Empire managed to establish a modicum of independence from Byzantium, and in 1077 their ruler Mihajlo (1051-81) was granted a royal crown in the pope's effort to gain a permanent foothold in the Orthodox East following the Great Schism of 1054. The Zeta Serb tribal kingdom, isolated in its mountainous environment, was of little consequence in the Byzantine Balkans. A truly influential Serbian state emerged in Raška under the reign of Stefan I Nemanja, who after 1180 managed to throw off direct Byzantine control (1190) and unite Zeta, northern Albania, and much of present-day eastern Serbia under his authority. A devout Orthodox Christian, Nemanja definitively planted his Serbian state in the East European Orthodox East. After abdicating in favor of his son Stefan II Nemanja (1196-1227), the true founder of the Nemanja dynasty, the former ruler retired first to a Serbian monastery and then to Mount Athos, where he joined another son, Rastko (known as St. Sava), in founding the large and influential Slav monastery of Hilandar.

The reign of Stefan II began amid conflict with his brother Vukan, prince of Zeta, whom the Hungarians, pushing southward into the Balkans, supported. Stefan was forced to flee to Bulgaria, where King Kaloyan supplied him with a Cuman army in return for territories around Belgrade and Niš. The struggle with the Hungarians ended in successful mediation by Stefan's saintly brother Sava, and Stefan assumed the throne of Serbia. He received a royal crown from the pope through a legate in 1217, thus acquiring the title of "First Crowned." But Sava managed to gain recognition of an autocephalous Serbian Orthodox archbishopric from the Greek patriarch, then in Nicæa. Sava, the first Serbian archbishop, then crowned Stefan Orthodox ruler of Serbia in 1219 with a crown sent by the Nicæan patriarch, an act that ended Catholic hopes to dominate the Serbs.

Stefan's successors during most of the 13th century proved weak rulers unable to thwart the territorial ambitions of Serbia's powerful neighbors. Bulgaria entrenched its hold over Serbian lands in the east, while Hungary held the region around Belgrade and established suzerainty over Bosnia. A strong secular and clerical Serbian aristocracy, which dominated the royal office, emerged during that time. Vague legal notions of succession and inheritance led to dynastic conflicts and regional disturbances exacerbated by continuing struggles between adherents of Catholicism and Orthodoxy, and by the presence of Bogomil heretics, particularly in Bosnia.

Stefan Uroš II Milutin (1282-1321), a pious yet dissolute person and an opportunist in religious and political matters, restored royal authority in Serbia. Taking full advantage of the growing weakness of the restored but truncated Byzantine Empire, Uroš gradually expanded Serbia into northern areas of Macedonia, along the Adriatic coast, and into Hungarian-held territories near Belgrade. His character was aptly demonstrated in his domestic life and in dealings with the Byzantines. He had a legal wife but maintained affairs with two known, officially kept concubines, as well as with a Greek princess from Thessaly. When his wife died (1297), Emperor Andronicus II proposed a marriage alliance with Uroš, which flattered the Serb king immensely, even though he was conducting a known affair with a nun, who was also his sister-in-law. The bride was five-year-old Princess Simonis, the emperor's daughter. (Uroš was then in his forties.) They were wed in Thessaloniki in 1299. Simonis was kept in the royal nursery for a few years before the lecherous Uroš consummated the marriage. Milutin's Macedonian conquests and marriage brought Serbia definitively into the Byzantine cultural orbit.

Uroš was succeeded by his illegitimate son Stefan Uroš III Dečanski (1321-31), who won a decisive victory over the Bulgarians and the Byzantines outside of Kyustendil (Velbuzhd) in 1330. Soon afterward, Byzantium disintegrated into civil war and Bulgaria was reduced to being Serbia's subordinate ally. (See Map 18.)

Dečanski's son Stefan Uroš IV Dušan brought Serbia to the pinnacle of its historical power and glory. He began his career by deposing his father and having him strangled. He came to rule over a Serbian state that included Raška, Zeta, Macedonia, Albania, Epiros, and Thessaly down to the Gulf of Corinth. He pushed the Hungarians north of the Danube and incorporated Belgrade and its environs into his large Balkan Serbian state. His attempts to conquer Bosnia proved unsuccessful, but Dušan cemented an alliance with the Bulgarians that spread Serbian influence into the eastern Balkans, and he tried to remain on fairly good terms with the Hungarians and Dubrovnik to give himself a free hand in exploiting the continuing civil disorders in the Byzantine Empire.

At his Macedonian capital of Skopje in 1346, Dušan had himself proclaimed emperor of the Serbs, Greeks (Byzantines), Bulgarians, and Albanians and was crowned as such by the archbishop of Peć, whom he then raised to the position of independent Serbian patriarch. A legal code for his Serbian "empire" was promulgated, and Dušan's court at Skopje took on all the outward trappings of Byzantine splendor. Before he could execute a planned advance on Constantinople and make good on his imperial pretensions, Dušan died. His death in 1355 was a catastrophe for the Orthodox Balkans, since it removed the last force capable of withstanding the advance into Southeastern Europe of the militant Islamic and expansionary Ottoman Turks. (See Map 19.) Soon after his death, Serbia fell into internal disarray, with local rulers throwing off the central authority of the weakened successors to the Serbian throne.

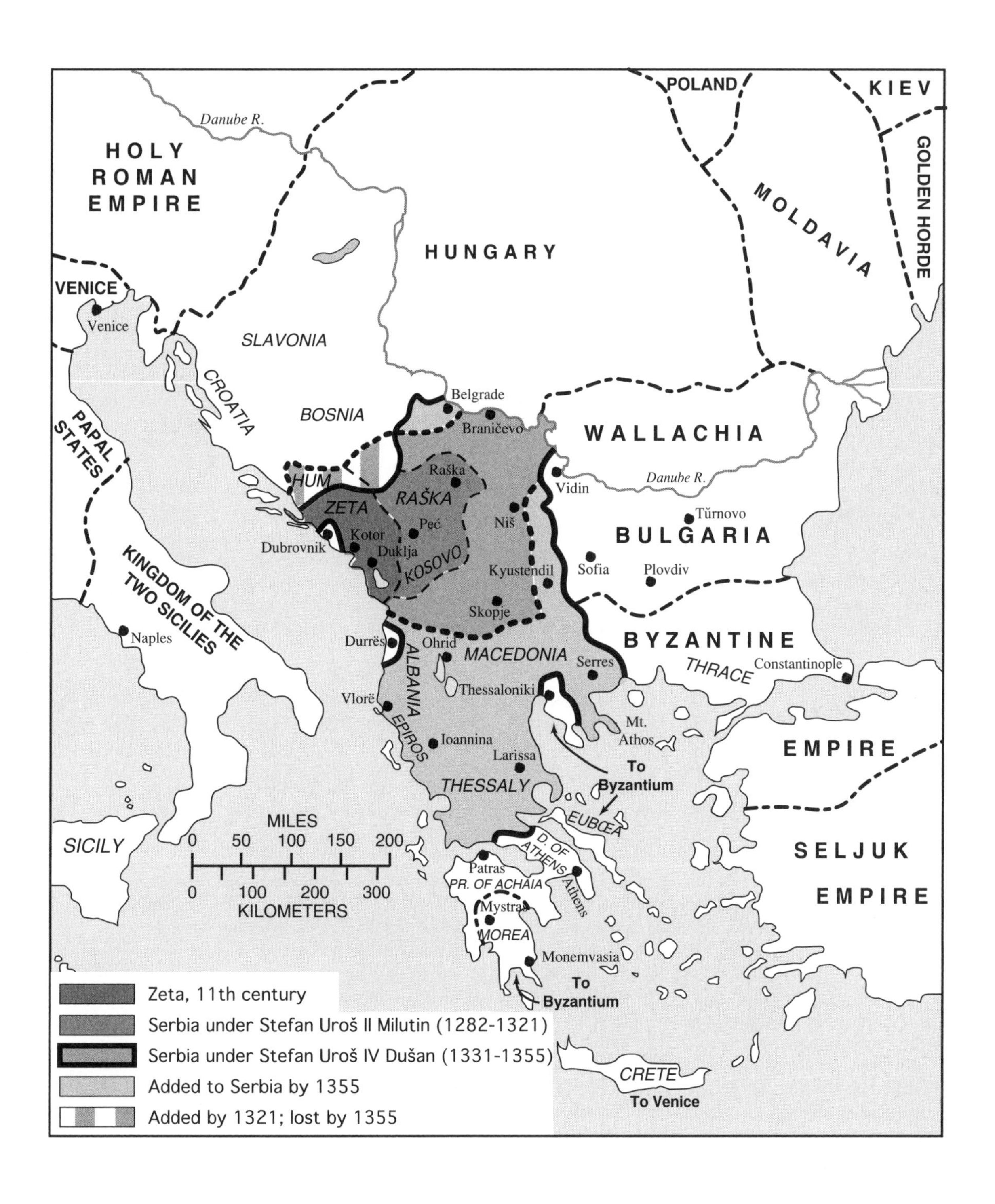

Danube R.
HOLY ROMAN EMPIRE
POLAND
KIEV
GOLDEN HORDE
MOLDAVIA
HUNGARY
VENICE
Venice
SLAVONIA
CROATIA
BOSNIA
PAPAL STATES
Belgrade
Braničevo
WALLACHIA
HUM
Raška
RAŠKA
ZETA
Vidin
Danube R.
Tŭrnovo
Niš
Peć
Kotor
Dubrovnik
Duklja
KOSOVO
BULGARIA
KINGDOM OF THE TWO SICILIES
Kyustendil
Sofia
Plovdiv
Skopje
Naples
Durrës
Ohrid
BYZANTINE
ALBANIA
MACEDONIA
Serres
THRACE
Constantinople
Thessaloniki
Vlorë
EPIROS
Mt. Athos
Ioannina
EMPIRE
Larissa
To Byzantium
THESSALY
EUBŒA
MILES
0 50 100 150 200
0 100 200 300
KILOMETERS
SICILY
D. OF ATHENS
Patras
PR. OF ACHAIA
Athens
SELJUK EMPIRE
Mystras
MOREA
Monemvasia
To Byzantium
CRETE
To Venice
Zeta, 11th century
Serbia under Stefan Uroš II Milutin (1282-1321)
Serbia under Stefan Uroš IV Dušan (1331-1355)
Added to Serbia by 1355
Added by 1321; lost by 1355

Map 18: Political Fragmentation, Mid-14th Century

Following a half century of Mongol-Tatar vassalage and disruptive civil wars among contestants for the throne, Teodor Svetoslav (1300-21) was installed as Bulgarian *tsar* in 1300. When Svetoslav died, the Bulgarian *bolyars* turned to Mihail Shishman, independent ruler of Vidin, as the strongest candidate capable of holding the throne in the face of possible Byzantine assault. Tsar Mihail Shishman (1323-30) reunited Vidin with Bulgaria and regained all territories lost in northern Thrace during recent Byzantine campaigns. In 1326 he concluded peace with Byzantium, which at the time was rent by civil war.

Emperor Andronikos II, an old and unpopular man, was overthrown in 1328 by his grandson and heir Andronikos after a desultory seven-year civil war. The new emperor, Andronikos III Palaiologos (1328-41), was young, frivolous, and irresponsible. He and Mihail Shishman forged a military alliance to expel the Serbs from Macedonia, and in 1330, a decisive battle was fought between Stefan Dečanski's Serbs and Mihail's Bulgarians near Kyustendil (Velbuzhd), in which the Bulgarian army was destroyed and Mihail fatally wounded. (See Map 17.)

The battle had crucial consequences. Bulgaria, under its new ruler Ivan Aleksandŭr (1331-71), was militarily crippled and politically subordinated to Serbia's interests. As for Byzantium, in just over a decade after Kyustendil the empire slipped into another civil war on Andronikos III's death. John V Palaiologos (1341-76) was a child when his father died, and the regency of John Kantakouzenos was opposed by John's mother and the patriarch. With both the Serbs and Bulgarians threatening the empire's borders, and Anatolian Turk raiders plundering Thrace, Kantakouzenos raised troops with his own money and checked them all. In his absence, the regency was taken over by his opponents. Kantakouzenos responded by declaring himself young John's coemperor and initiated a war against his enemies in the capital.

The new civil war became bound up with radical social reform (Zealot) and religious (Hesychist) movements, with the Zealots opposed to and the Hesychists supportive of Kantakouzenos. By 1342 Kantakouzenos's fortunes were in decline and he appealed to Serbian ruler Dušan for help. When, in the next year, the governor of Thessaly recognized Kantakouzenos as emperor, Dušan dropped him and joined forces with the regency in Constantinople. Kantakouzenos then called on assistance from the Seljuk and Ottoman Turks in Anatolia, whose combat and plundering in Thrace tipped the military scales in his favor. In 1347 Kantakouzenos entered Constantinople as Emperor John VI (1347-54), and young John V was reduced to junior imperial partner.

Bulgarian Tsar Ivan Aleksandŭr had intervened on the regency's side during the civil war, and Kantakouzenos had unleashed his Ottoman allies on him. Thereafter, paralyzing Turkish raids on Bulgaria became commonplace and Bulgaria rapidly slid into decline. To consolidate his authority, Aleksandŭr divided the state into three appanages held by his sons, but that arrangement proved only partially successful. In the 1340s, a *bolyar* named Balik tore Dobrudzha from Aleksandŭr's control, and during the 1350s Bulgaria was rent by a half-brotherly rivalry between the holder of Vidin, Ivan Stratsimir, and Aleksandŭr's designated successor and holder of Tŭrnovo, Ivan Shishman. Stratsimir proclaimed himself *tsar* of an "empire" of Vidin in 1370, and Dobrotitsa (ca. 1366-85), Balik's successor (after whom Dobrudzha became named), received de facto recognition as independent despot in Dobrudzha. Political disunity was magnified in the religious sphere by the ascendancy of intolerant Hesychism within the Bulgarian Tŭrnovo Orthodox Patriarchate, which caused popular unrest and a wave of emigrations precisely when the Ottoman threat to Bulgaria reached serious proportions. (See Map 19.)

The internal chaos that wracked both Byzantium and Bulgaria permitted Dušan to carve out his large Serbian Empire at their expense. (See Map 17.) Dušan died in late 1355 before he could march on Constantinople and attempt to finish off the stricken Byzantine Empire. Into the grave with him went his empire. Stefan Uroš V (1355-71), young, weak, and perhaps mentally handicapped, lacked his father's talents and forceful personality needed to hold the disparate elements of the empire together. Released from Dušan's control, the powerful regional Serb nobles rapidly turned their provincial holdings into fully or semi-independent principalities, retaining only the most nominal ties to Dušan's impotent successor. Serbia fragmented.

First to throw off Serbian control were the Greek provinces of Thessaly and Epiros, as well as Dušan's former Albanian holdings. A series of small independent principalities arose in western and southern Macedonia, while the Hungarians encroached deeper into Serb lands in the north. Uroš held only the core Serbian lands, whose nobles, though more powerful than their ruler, generally remained loyal. These core lands consisted of the western lands, including Montenegro (Zeta); the southern lands, held by Jovan Uglješa in Serres, encompassing all of eastern Macedonia; and the central Serbian lands, stretching from the Danube south into central Macedonia, co-ruled by Uroš and the powerful noble Vukašin Mrnjavčević, who held Prilep in Macedonia. Instead of preserving Serb unity, Uroš's loosely amalgamated domains were wracked by constant civil war among the regional nobles.

In 1354 the Byzantine fortress of Gallipoli was ruined by an earthquake and occupied by Ottoman troops serving Kantakouzenos. They quickly established Gallipoli as a permanent European base for raiding operations into Thrace. Nearby Constantinople was thrown into panic and the usurper, Kantakouzenos, was blamed and easily toppled by John V Palaiologos. Despite the return of legitimate rule, Byzantium was ruined. No strong state existed to protect the Balkans from the Ottoman Turks' impending onslaught. (See Map 19.)

HOLY ROMAN EMPIRE
HUNGARY
KIEV
BESSARABIA
GOLDEN HORDE
MOLDAVIA
Danube R.
Bratislava
Vienna
Suceava
VENICE
Zagreb
ISTRIA
CROATIA
TRANSYLVANIA
SLAVONIA
Belgrade
Braničevo
Tîrgovişte
WALLACHIA
Visoko
PAPAL STATES
Zadar
Split
BOSNIA
Vidin
Danube R.
Silistra
DOBRUDZHA DESPOTATE
VIDIN EMPIRE
SERBIA
Niš
Tŭrnovo
Kavarna
Varna
KINGDOM OF THE TWO SICILIES
Dubrovnik
Peć
KOSOVO
II
Sofia
BULGARIA
TŬRNOVO EMPIRE
Shkodër
Kyustendil
Plovdiv
I
Prilep
III
Adrianople
Naples
Durrës
Ohrid
MACEDONIA
V
IV
Serres
Constantinople
THRACE
EPIROS
VII
Thessaloniki
To Sicily
VI
Bursa
Butrint
Larissa
OTTOMAN EMPIRE
Arta
CORFU
NEOPATRAS
D. OF ATHENS
EUBŒA
SELJUK EMPIRE
SICILY
Patras
PR. OF ACHAIA
Athens
Mystras
MOREA
Monemvasia
MILES
0 50 100 150 200
0 100 200 300
KILOMETERS
CRETE
Byzantine Empire after 1355
Bulgaria before 1357
Serbia after 1355
Venetian possessions
Regional Successor Principalities after 1355
I: of Vukašin and Karalj Marko
II: of Velbuzhd (Kyustendil)
III: of Hrelyo
IV: of the Bogdanovs
V: of Uglješa
VI: of Simeon
VII: of the Hlapenovs

Era of Ottoman Domination
1355-1804

Map 19: Ottoman Expansion in the Balkans, Mid-14th–Early 16th Centuries

The term "Ottoman" is a Western corruption of the Turkish name of the original tribal leader Osman I (1281-1324). He ruled the Seljuk principality closest to Byzantium and Europe in the northwest corner of Anatolia and pursued unrelenting warfare against the Christians directly across his borders. This soon attracted to him a number of warriors—organized into an effective and loyal military force—from all parts of the Seljuk world eager to expand Islamic territories in the tradition of the *jihad* (Holy War). After his death, Bursa was captured and the Byzantines were completely expelled from Anatolia by his son and immediate successor Orhan I (1324-60). Under Orhan, the Ottomans (the collective name for the assorted warriors and allies of the house of Osman) permanently established themselves in Southeastern Europe. For the next two and a half centuries, their string of military successes against European Christians stretched unbroken under ten consecutive rulers.

Ottoman forces first entered Europe in 1345 as mercenary allies of John VI Kantakouzenos in his civil war with Emperor John V Palaiologos. In 1349 Orhan again sent military support to John VI to counter Dušan's Serbian encroachments. When John called on Turkish help for the third time in 1354, Orhan's forces did not return to Anatolia as they had previously, but took control of Gallipoli in Europe after it was damaged by an earthquake, fortified it, and transformed it into a permanent base for expansionary operations in the Balkans. By Orhan's death, the Ottomans were entrenched in Europe, their state was well-organized, and the Byzantine Empire was at their mercy.

Ottoman sultan Murad I (1360-89) realized that the Balkan Christian states of Byzantium, Serbia, and Bulgaria had been weakened by decades of internecine wars. Their populations were burdened with rising semifeudal oppression, economic disruption, and unstable living conditions. Murad captured Adrianople in 1365 and transformed it into his European capital Edirne. A new standing professional infantry force called the Janissaries was founded to supplement the traditional Turkish tribal cavalry units. The new troops consisted first of enslaved war captives and later of child-levies from among the growing numbers of Balkan subject Christians. Murad conquered the Bulgarian lands south of the Balkan Mountains by 1372 and reduced its Tŭrnovo ruler Tsar Ivan Shishman to vassal status. In 1371 he destroyed a predominantly Serbian force, composed of local Macedonian regional troops led by Jovan Uglješa of Serres and Vukašin Mrnjavčević of Prilep, at Ormenion (Chernomen) outside of Edirne, an act that led to the conquest of Macedonia. By 1386 Murad had taken the regions around Sofia and Niš and forced a weakened Serb ruler Prince Lazar (1371-89) into submission as his vassal.

As Murad's state expanded, it became clear that the Ottomans were in Europe to stay. Many Christian rulers, such as Shishman and Lazar, as well as their independent-minded warrior nobility, joined the Ottomans as allies in an effort to retain their political and social positions. Christian Bulgarians and Serbs, including Vukašin's son, the legendary Kralj Marko of Prilep, fought loyally in the Ottoman ranks throughout most of the campaigns that won control of the Balkans, including that which resulted in the Battle of Kosovo Polje (1389), when Serbia definitively was broken. Lazar had joined a coalition of Serb, Bosnian, Albanian, and Wallachian magnates in an effort to contain the Ottomans. Murad set out to punish his unfaithful vassal. After the battle, in which both Lazar and Murad were killed, Serbia's fragmentation accelerated.

Murad's death did little to stop the Turkish onslaught. Bayezid I the Thunderbolt (1389-1402) continued Ottoman expansion in rapid campaigns that made Serbia a vassal state, incorporated Bulgaria outright (1393), and reduced Wallachia to subordinate vassal status. Constantinople was besieged unsuccessfully (1391-98) and much of Greece conquered. A Western crusade against the Ottomans, led by Hungarian king Sigismund (1387-1437), was destroyed near Nikopol on the Danube in 1396. Ottoman European momentum was stymied, however, by the Mongol invasion of Anatolia (1402), which Bayezid proved powerless to stop. During the interregnum (1402-13) following his death, Ottoman expansion in the Balkans temporarily stalled and Serbia was able to reassert a modicum of independent action.

The brief respite for the Balkan Christians ended when the Ottomans returned in renewed force under Mehmed I the Restorer (1413-21), who reasserted his authority over most of the Ottomans' European possessions. His son Murad II (1421-51) resumed Ottoman expansion. A war with Venice resulted in the conquest of Thessaloniki (1430) and most of the Aegean islands. Hungary was unsuccessfully invaded (1442), and the subsequent and final Hungarian-led crusade against Islam, commanded by the renowned general János Hunyadi, was crushed outside the Bulgarian seaport of Varna (1444). At Murad's death, little remained of the Byzantine Empire other than Constantinople itself.

Mehmed II the Conqueror (1451-81) captured Constantinople in 1453. The city was renamed Istanbul and transformed into the capital of the Islamic Ottoman Empire, thus continuing its tradition as the center for a divinely ordained world state. (See Map 20.)

Once established in his new capital, Mehmed set out to complete the total conquest of the Balkans and to push the borders of his empire deeper into Europe. Serbia finally was subdued and incorporated (1456-58); Bosnia and Hercegovina conquered (1458-61); the tough resistance of the Albanians, led by George Kastriotis (Skanderbeg), broken (1456-63); and Venice's presence in the eastern Mediterranean reduced (1463-79).

POLAND
HOLY ROMAN EMPIRE
Danube R.
Vienna
Buda
Cluj
HUNGARY
TRANSYLVANIA
Suceava
BESSARABIA
MOLDAVIA (Vassal Client)
Cetatea Albă
VENICE
Venice
Zagreb
CROATIA
SLAVONIA
Belgrade
Smederevo
Tirgoviște
Bucharest
WALLACHIA (Vassal Client)
DOBRUDZHA
PAPAL STATES
Zadar
DALMATIA
BOSNIA
Sarajevo
SERBIA
Vidin
Nikopol
Danube R.
BULGARIA
Tŭrnovo
Varna
HERCEGOVINA
MONTENEGRO
Niš
Kosovo Polje
Dubrovnik
Sofia
Skopje
Lech
ALBANIA
Plovdiv
Adrianople
Naples
NAPLES
Durrës
Ohrid
MACEDONIA
THRACE
Constantinople
Thessaloniki
Gallipoli
Ioannina
EPIROS
Larissa
THESSALY
Bursa
BYZANTINE EMPIRE
MILES
0 50 100 150 200
0 100 200 300
KILOMETERS
Navpaktos
EUBOEA
SELJUK PRINCIPALITIES
K. of SICILY
Athens
(To Genoa)
Mystras
MOREA
Methoni
Monemvasia
RHODES (To Knights of St. John)
(To Venice)
CRETE
Venetian possessions
Ottoman possessions,1326
Territories acquired by 1360
Territories acquired by 1389
Territories acquired by 1402
Territories acquired by 1481
Territories reduced to vassal clientage by 1402, permanently annexed by 1481
Territories acquired by 1504

Map 20: Fall of Constantinople, 1453 (and Ottoman Istanbul)

Mehmed II assumed the Ottoman throne determined to capture the Byzantine imperial city of Constantinople and transform it into the capital of an Islamic Ottoman Empire. After cutting the city off from all outside assistance, Mehmed laid siege to Constantinople in 1453, conveying part of his navy overland from the Bosphorus to the Golden Horn to completely isolate the city from any last-minute outside reinforcement. His heavy artillery, purchased from the Hungarian gun maker Urban, blasted holes in the once impregnable land walls, and on 29 May, the final vestige of the Christian Roman Empire fell, with its last emperor, Constantine XI (1448-53), dying heroically defending the walls.

After three days of obligatory sacking, as the city had been captured by force, Mehmed began embellishing his new capital. Justinian's cathedral of Hagia Sophia was converted to an imperial mosque, as eventually were other churches and monasteries. The rights of non-Turkish inhabitants were protected to ensure continuity and stability for commercial activities. Because Constantinople had never fully recovered from the sack of 1204, and because after 1261 the restored Byzantine Empire existed in a state of near-poverty, by the time of the conquest Constantinople had become a hollow shell of its former self. Its population had dwindled and much property was either abandoned or in a state of disrepair.

Mehmed envisioned Istanbul as the capital of a powerful, highly cultured, civilized Ottoman Islamic world state representing the divinely ordained order for all humankind on earth (a view similar to the traditional one of the Orthodox Christians regarding Byzantium). In many respects, Istanbul was to continue Constantinopolitan tradition as the political and cultural fountainhead for an essentially theocratic society. Formerly, that society had been Christian Byzantium. Now it was to be the Islamic Ottoman Empire.

The Ottomans quickly began repopulating the city. Mehmed offered civic and private properties to the public to entice skilled artisans, craftsmen, and traders, of all religions and ethnicities, to reside within the city's walls. Istanbul rapidly grew into a multiethnic, multicultured, and bustling economic, political, and cultural center for the Ottoman state, whose distant frontiers guaranteed it peace and security.

Within its ancient walls, Istanbul steadily acquired a distinctive Islamic character. First rose numerous minarets added to former Christian churches converted into mosques. Then came a plethora of new structures: Mosque complexes, fountains, caravansaries, public baths, public soup kitchens and hospices, tombs and mausoleums, dervish convents, libraries, and other such edifices. The old Byzantine palaces having fallen into disrepair or ruin, Mehmed first built a new one for himself (Eski Sarayı), scavenging materials from the older, dilapidated Church of the Holy Apostles. He soon decided (in 1459) to construct a larger one on the crest of Byzantion's old acropolis, in the areas of the former Byzantine Great and Mangana palaces. Surrounded by massive defense walls and encircled by extensive parks and gardens, Topkapı, which eventually encompassed a complex series of small, individual private buildings (bedrooms, harem, libraries, and kiosks) and functional ones (divan, treasury, reception halls, kitchens, guard barracks, and arsenal), became the political nerve center of the empire. One of the gates in the palace walls, through which passed all foreign plenipotentiaries to the imperial court, eventually gave the Ottoman government its common identity as the "Sublime Porte."

All political, military, religious, and cultural life in Istanbul revolved around Topkapı and nearby Aya Sofya. Here were stationed the Janissaries; public spectacles were conducted in the Hippodrome; important religious and political functionaries built palaces in the area. Much of the loot from successful military campaigns, and a significant percentage of tribute and taxes from conquered territories, was used to adorn Istanbul with beautiful mosques, palaces, and pious foundations. No sultan was more active in such activity than Süleyman I the Magnificent (1520-66), whose imperial architect Sinan gave the city some of its most exceptional masterpieces. The Süleymaniye mosque complex (1557) by Sinan, which dominated the city's skyline, was a symbol of the power and glory of the Ottoman Empire at its height, serving essentially the same purpose as did Hagia Sophia for Byzantium.

The inhabitants of Istanbul were a patchwork of ethnicities and religions, who often lived in city quarters identified by a particular ethnicity. There were Jewish and Armenian quarters, for example. The most significant quarter, as far as the Balkans were concerned, was the Phanar (lighthouse) district on the south bank of the Golden Horn near the northern terminus of the old land walls. This was the quarter inhabited by wealthy Greek merchants, and where the Greek Orthodox patriarchate was eventually established. The Greeks of the quarter, known as Phanariotes, played important economic and political roles in the Ottoman Empire in the 16th through 18th centuries. They dominated the empire's international maritime and overland commercial activities; commanded the empire's navy; staffed the imperial office for foreign affairs as administrators and translators (because of their extensive foreign contacts); and purchased the crowns of the tributary Wallachian and Moldavian principalities from the Ottomans throughout the 18th century. Most important, the Phanariotes controlled the office of Greek patriarch, which, because of the Ottoman *millet* system, gave them authority over all Orthodox Christians, no matter their ethnicity, within the empire. As Orthodox Christians represented the single most significant source of imperial tax revenues, the Phanariotes became influential in Ottoman affairs. (See Map 22.)

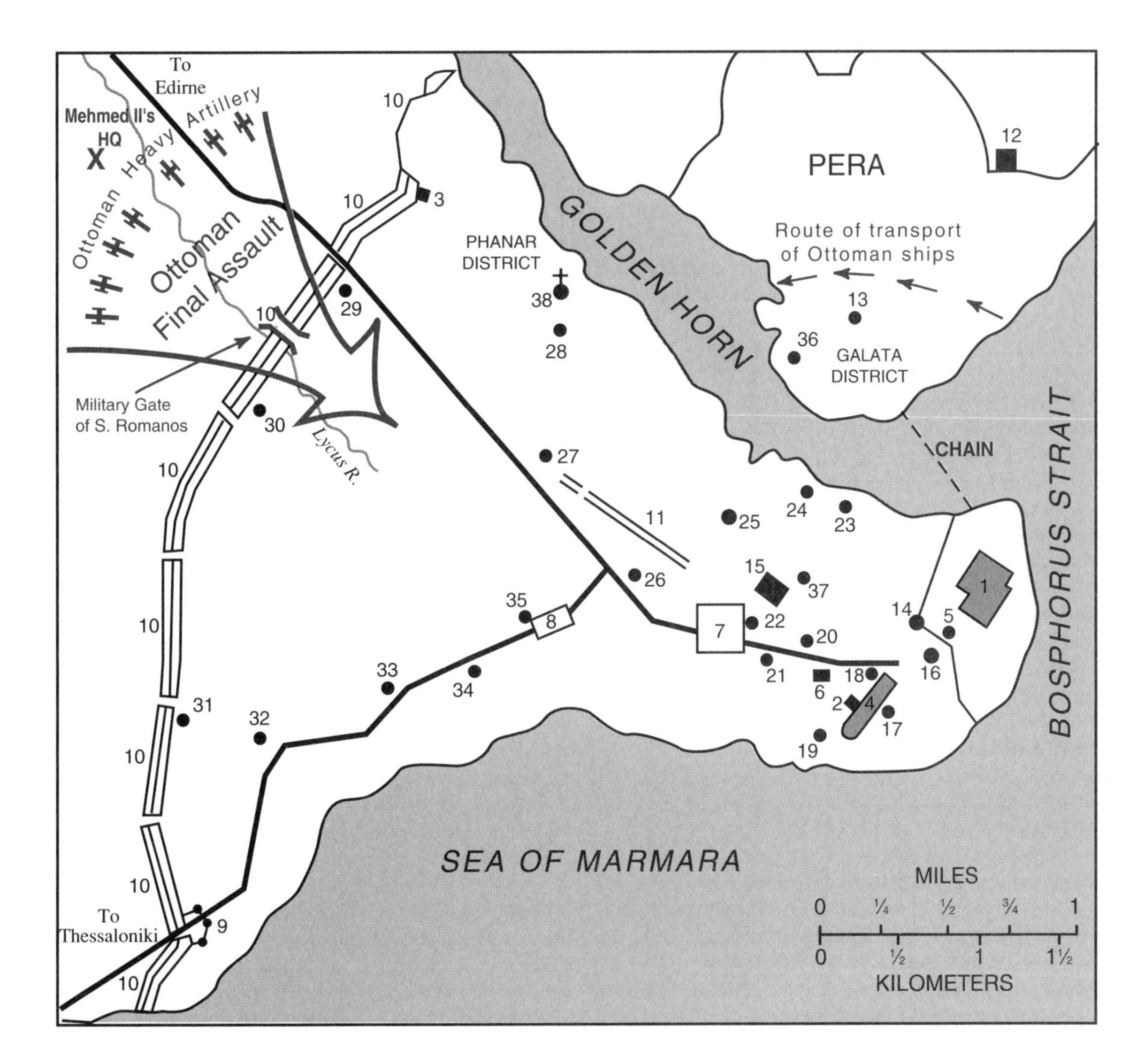

Palaces

1. Topkapı
2. Ibrahim *paşa*
3. Tekfur Saray

Civic Structures

4. Hippodrome
5. Janissary Arsenal (Old St. Irene)
6. Köprülü Complex
7. Bayezid Square
8. Aksaray Square
9. Seven Towers Fortress
10. Land Walls
11. Aqueduct of Valens
12. Tophane Cannon Foundry
13. Galata Tower
14. Sublime Porte
15. Covered Bazaar

Mosques and Mosque Complexes

16. Aya Sofya (Old Hagia Sofia)
17. Ahmed I (Blue Mosque)
18. Firuz *ağa*
19. Sokollu Mehmed *paşa*
20. Atik Ali *paşa*
21. Kara Mustafa *paşa*
22. Bayezid II
23. Yeni Valide
24. Rüstem *paşa*
25. Süleymaniye
26. Şehzade
27. *Fatih* Mehmed II
28. Selim I
29. Mihrimah
30. Kara Ahmed *paşa*
31. Ibrahim *paşa*
32. Ramazan *efendi*
33. Isa Kapı
34. Davut *paşa*
35. Haseki Hürrem
36. Azap Kapı
37. Mahmud *paşa*

Church

38. Greek Orthodox Patriarchate

Map 21: Apex of Ottoman Expansion, Mid-16th Century

The reign of Sultan Süleyman I the Magnificent represented the high-water mark of Ottoman expansion in Europe. Süleyman's military forces, anchored by his highly disciplined corps of household-slave Janissary infantry and guard cavalry, were victorious in most battles with their European enemies. Military successes, and the extension of his empire deep into the Danubian Basin, made Süleyman preeminent among European rulers, and his reign marked the "golden age" of Ottoman architecture, fine arts, law, literature, diplomacy, and commerce.

Son of Sultan Selim I the Grim (1513-20), who spent his reign focused on Ottoman expansion in West Asia and North Africa, Süleyman turned his attentions to Europe. In 1521 he captured Belgrade, a key fortress guarding Hungary's southern border. Intermittent border warfare with the Hungarians culminated in the decisive Battle of Mohács (1526), in which the Hungarians were crushed and their king Louis II Jagiełło (1516-26) killed. Mohács opened the way to further Ottoman advances into the heart of Central Europe. The Hungarian capital of Buda was occupied, and Süleyman, deciding at first not to annex Hungary, made it a tributary state under his Transylvanian vassal János Zápolya (1526-40). This triggered a civil war within Hungary pitting Zápolya against the Habsburg contender for the throne, Ferdinand I (1526-64). When in 1527 Ferdinand's forces captured Buda and defeated Zápolya in battle, the Transylvanian appealed to Süleyman for help.

Süleyman prepared a new expedition into Hungary and concluded a secret anti-Habsburg alliance with France (the first between a European Great Power and the Ottomans). In May 1529 he led his forces north of the Danube. Buda was recaptured after a short siege in September and Süleyman pushed into the Habsburg Empire, but stubborn Habsburg resistance and approaching winter weather frustrated his attempt to take besieged Vienna and forced him to withdraw. Ottoman-Habsburg fighting continued in the north of Hungary until 1533, at which time a war with Persia in West Asia forced Süleyman to sign a treaty with Ferdinand recognizing Habsburg control of a strip of northern Hungary-Croatia, while Zápolya retained his hold on the remaining two-thirds of that kingdom. Both were made to pay tribute to Süleyman for their Hungarian possessions.

Ferdinand resumed warfare against Süleyman in 1537 as part of a joint anti-Ottoman and anti-French alliance with the Holy Roman Empire, the Papal States, and Venice organized by his brother, Holy Roman emperor Charles V (1519-56). In 1541, with Habsburg forces consistently unsuccessful and with the death of Zápolya (1540), Süleyman decided to annex Zápolya's portion of central Hungary. Three years of inconclusive campaigning followed. Events in West Asia made it necessary for Süleyman to renew the war with Persia, so he felt constrained to make peace with the Habsburgs on the European frontier. In 1544 he made another agreement with Ferdinand based on the status quo ante. This permitted Ferdinand to hold his strip of Royal Hungary in return for continued annual tribute payments.

A short lull in the Ottoman-Habsburg conflict followed the signing of the peace agreement. Transylvania remained a tributary vassal state of the Ottomans under a native prince. The Romanian principalities of Wallachia and Moldavia also were governed by native princes who were tribute-paying vassals of the sultan. While there was no Ottoman military presence in Transylvania, the Ottomans maintained a few garrisons in fortresses built on Romanian territories to protect the empire's Danubian and Ukrainian defenses. All three vassal states were required to pay the Ottomans an annual tribute, their ruling princes could not assume power without the confirmation of the Ottoman sultan, and certain trade commodities (especially foodstuffs) had to be sent to the empire. In return, the Ottomans permitted them internal political, social, and cultural autonomy. So long as the vassal states met their monetary and service obligations and did not act on the international scene in ways deemed detrimental to Ottoman foreign policy, the Ottomans were content to let the native princes govern without much interference. Elsewhere in the Balkans, the Dalmatian Dubrovnik Republic technically was an Ottoman vassal, while only Habsburg Croatia and Venetian Dalmatia lay outside of the sultan's direct authority (although his rule over mountainous Montenegro was little more than nominal).

In 1551 Ferdinand renewed the war against the Ottomans by invading Transylvania, which he successfully held for two years before being repulsed. Following Charles V's abdication in 1556, Ferdinand was elected Holy Roman emperor (1556-64) and the war with the Ottomans settled into desultory border fighting. Süleyman and Ferdinand brought the intermittent fighting to a close with the Peace of Prague (1562), with no real change in the status quo. Ferdinand continued to pay Süleyman tribute for his Hungarian holdings until his death in 1564. When Ferdinand's imperial successor Emperor Maximilian II (1564-76) ordered renewed raids against the Ottomans in 1566, Süleyman, then 72 years old and suffering from gout, led an army against Royal Hungary and laid siege to the fortress of Szigetvar. There he died in his tent, two days before the citadel fell to his troops, who were prevented from learning of his death until after their victory.

Although the Ottomans would continue to pose a serious threat to Western Europe for another century, and would even gain temporary new conquests in southern Poland and Ukraine, their fortunes in Europe crested with the life and death of Süleyman. The decline began during the reign of his son, whose descriptive title aptly fit his character, Sultan Selim II the Sot (1566-74).

POLAND
HABSBURG EMPIRE
ROYAL HUNGARY
HUNGARY
Danube R.
Vienna
Bratislava
Buda
Pest
Suceava
MOLDAVIA
(Vassal Client)
BESSARABIA
Cluj
TRANSYLVANIA
(Loose Vassal Client)
Cetatea Alba
Zagreb
CROATIAN MILITARY BORDER
Venice
Szigetvar
Mohács
SLAVONIA
CROATIA
Belgrade
WALLACHIA
(Vassal Client)
DOBRUDZHA
PAPAL STATES
Zadar
DALMATIA
Split
BOSNIA
Sarajevo
SERBIA
Bucharest
Danube R.
Niš
OTTOMAN EMPIRE
MONTENEGRO
Dubrovnik
(Client Vassal)
Varna
BULGARIA
KOSOVO
Sofia
Plovdiv
Skopje
Edirne
Naples
KINGDOM OF NAPLES
Durrës
ALBANIA
Ohrid
MACEDONIA
THRACE
Istanbul
Thessaloniki
Ioannina
EPIROS
Larissa
THESSALY
CORFU
Navpaktos
Athens
MOREA
K. of SICILY
MILES
0 50 100 150 200
0 100 200 300
KILOMETERS
Methoni
Monemvasia
RHODES
CRETE
Ottoman border, mid-16th century
Territories acquired between 1505 and 1566
Ottoman vassal client states
Venetian possessions

Map 22: Ottoman Millet Organization, Mid-16th–17th Centuries

Until the early 16th century and the expansionary campaigns of Selim I in Islamic West Asia and Egypt, the Ottomans' non-Muslim Balkan subjects equalled or outnumbered the Muslims in the Ottoman Empire as a whole. The empire's numerous Balkan non-Muslim subjects posed legal and administrative difficulties for the governing authorities because, in an Islamic state, the *Şeriat* (Sacred Law) was applicable only to Muslims and possessed no validity among non-Muslims. By the time of Constantinople's conquest, close to half of the empire's subject population lay outside of the law. Mehmed II found it necessary to create an institutional structure for administratively integrating non-Muslim subjects into the theocratic Ottoman state. In 1454 he instituted the *millet* system of non-Muslim administration.

Reasoning that the various non-Muslim groups' religious laws could govern them as the *Şeriat* did Muslims, Mehmed divided his subjects into *millets* (nations) based solely on religious affiliation and administered by the highest religious authorities of each. All non-Muslims were distributed among three *millets* encompassing the most important existing non-Muslim faiths: Orthodox Christians, headed by the patriarch of Constantinople and representing the single largest and economically most important non-Muslim group; Jews, of great commercial significance, headed by an elected official of the rabbinical council in Istanbul; and Armenian (Gregorian Monophysite) Christians, headed by an Armenian patriarch of Istanbul appointed by the sultan, which also included the empire's Roman Catholic subjects. Each *millet* represented its membership before the Ottoman court and was internally self-governing. They all were granted the rights to tax, judge, and order the lives of their members insofar as those rights did not conflict with Islamic sacred law and the sensibilities of the ruling Muslims. The religious hierarchies of the *millets* were endowed with civil responsibilities beyond their ecclesiastical duties, and their head prelate was held accountable for their proper internal functioning. In effect, each *millet* became an integral part of the empire's domestic administration, functioning as a veritable department of the Ottoman central government. In return for ensuring the smooth administration of their non-Muslim subjects, the Ottomans granted the *millets* considerable autonomy in the spheres of religious devotion and cultural activity, judicial affairs not involving Muslims, and local self-government.

While the Turkish term *millet* involved the idea of "nation," it shared little with the Western European concept. *Millet* identified people solely on the basis of religion; ethnicity and territory played no role. Although the Muslim authorities were aware of ethnic differences, given their theocratic mentality, ethnicity was thought relatively unimportant. Until the close of the 18th century, *millet* affiliation—religious belief—was the fundamental source of group identity among all of the empire's Balkan subjects. Such identity, however, was not uniform since every *millet* represented a culturally diverse membership. That was crucial for the most important *millet*—the Orthodox.

The Orthodox *millet* was divided along linguistic lines between Greek- and Slavic-rite members. From inception, it was dominated by Greek speakers favored by the Ottomans, who, during their Balkan conquests, placed the Bulgarians' and Serbs' independent Slavic-rite Orthodox churches into the hands of the Greek Patriarchate of Constantinople. When the *millet* system was devised, the Greek patriarch was installed as head of the Orthodox *millet*, effectively placing its administration under Greek control.

Because its membership outnumbered all other non-Muslims, the Orthodox *millet* ranked first in Ottoman interests. The Greek church of Constantinople enjoyed administrative jurisdiction over the empire's entire Orthodox population, including the Slavic-rite Bulgarians and Serbs, who had possessed their own autonomous churches prior to the Ottoman conquest. One of the two former churches in Bulgaria, that at Ohrid, was reduced to an autocephalous archbishopric, officially retaining the label "Bulgarian" but headed predominantly by Greek prelates. The Patriarchate of Tŭrnovo was abolished outright following the Ottoman conquest of Bulgaria and its ecclesiastical jurisdiction placed under the Greek patriarch's direct authority.

As for the Serbian Peć Patriarchate, after Serbia's fall (1459) most of its territorial jurisdiction was taken over by the Ohrid archbishopric until the influential Ottoman grand vezir Mehmed Sokollu convinced Süleyman I to reinstate the autocephalous Serbian patriarchate in 1557. Because of favor in high Ottoman quarters, the Peć Patriarchate was staffed by Slavic-speaking clergy, although the Greek Patriarchate continuously exerted pressure on the Ottoman authorities to reduce its autonomy. Only within the confines of its jurisdiction did Slavic-speaking Orthodox believers enjoy the luxury of widespread Slavic cultural expression through official church support.

In regions directly under the Greek patriarch's control, the church hierarchy generally was dominated by Greek speakers who officially supported Greek cultural activities with church funds. Non-Greeks were relegated to the lowest rungs of the hierarchy and maintained non-Greek (Slavic) cultural life from the limited funds available through grassroots popular support, thus restricting the quantity and quality of their cultural endeavors. The cultural situation for Slavic speakers within the Ohrid archbishopric was only marginally better. So overwhelming was the Greek presence in the Orthodox *millet* that a perceptual association generally linking "Greek" and "Orthodox" became common among the Ottoman authorities and European travelers in the Balkans. Until well into the 18th century, all Balkan Orthodox believers were identified as "Greeks" by European observers.

In the mid-18th century the Greek Patriarchate, playing on political and socioeconomic circumstances holding in the Ottoman Empire at the time (see Map 24), convinced the Ottomans to abolish both the Peć (1766) and Ohrid (1767) churches and place their jurisdictions under direct Greek patriarchal control.

POLAND
HABSBURG EMPIRE
ROYAL HUNGARY
HUNGARY
Danube R.
Vienna
Bratislava
Buda
Pest
Cluj
Iaşi
BESSARABIA
MOLDAVIA
(Vassal Client)
TRANSYLVANIA
(Loose Vassal Client)
Kalocsa
Zagreb
CROATIAN MILITARY BORDER
Venice
ISTRIA
Sremsi Karlovci
SLAVONIA
CROATIA
Belgrade
WALLACHIA
(Vassal Client)
DOBRUDZHA
PAPAL STATES
Zadar
DALMATIA
BOSNIA
SERBIA
Bucharest
Split
Sarajevo
Niš
Turnovo
Varna
MONTE-NEGRO
Peć
OTTOMAN
BULGARIA
Dubrovnik
(Vassal Client)
Cetinje
KOSOVO
Sofia
Skopje
Plovdiv
Naples
KINGDOM OF NAPLES
Durrës
ALBANIA
MACEDONIA
Ohrid
Edirne
THRACE
Istanbul
Thessaloniki
EPIROS
Ioannina
Larissa
THESSALY
CORFU
EMPIRE
K. of SICILY
MILES
0 50 100 150 200
0 100 200 300
KILOMETERS
Athens
Patras
MOREA
CRETE
RHODES
Venetian Possessions
Orthodox Patriarchal seats
Southern boundary of Roman Catholic jurisdictions
Ottoman Border, mid-16th Century
Territories of Pec Slavic-Rite Patriarchate, 1557
Territories of Ohrid Slavic-Rite Archbishop-Patriarchate, 1557
Territories of Constantinople Greek Rite Patriarchate, 1557

Map 23: Hapsburg Croatian-Slavonian Military Border, 17th–18th Centuries

Since accepting dynastic political union with Hungary in 1102, the Croat nobility enjoyed "nation" status with the Hungarian kings. (See Map 10.) The partnership with Hungary was not always harmonious, but until the early 19th century, neither was it antagonistic. Under mostly loose Hungarian oversight, the Croat nobility exercised local autonomy within the territories of its historic kingdom, and many of its members came to play roles of importance within Hungary itself. In a disputed royal election after Hungary's defeat by the Ottomans at Mohács (1526), the Croat nobles supported the Habsburg candidate Ferdinand I on condition that he protect Croatia against the Ottomans and recognize their continued rights and privileges. Thereafter, they ranked among the Habsburgs' most loyal supporters in rump Royal Hungary, which primarily consisted of Croatia, a strip of northern Hungary, and Slovakia—all that remained of the former Hungarian Kingdom following Süleyman I's conquests in the 1540s. (See Map 21.)

Repeated Ottoman incursions into Royal Hungary from Bosnia resulted in the Habsburgs establishing the Croatian Military Border in 1538. Manned by settled military colonists commanded by the local Croat aristocracy, the border zone *(Krajina)* was placed directly under the authority of Habsburg military headquarters in Vienna. Only the northern Croatian region of Zagreb officially remained bound to Hungary. The Habsburg creation of the military border probably was inspired by a similar frontier organization founded by the Ottomans earlier in the 16th century in the northern Bosnian lands under their control.

The military border freed the Croat nobility from direct Hungarian control through the 17th century. There followed, however, a period of disappointment and growing discontent in the 18th century, during which the Habsburgs courted the Hungarians' support against threats from Prussia and other European rivals by gradually permitting them to reassert some authority in military Croatia.

When the Habsburgs acquired Slavonia in the late 17th century, it was designated a military border region similar to the Croatian one, and the border system was extended into Vojvodina by Emperor Leopold I (1658-1705) following the Treaty of Sremski Karlovci (1699). (See Map 24.) Needing to maintain the frontier troops' numerical strength, the Habsburgs settled an ethnic hodgepodge of available peoples as military colonists—Croats, Germans, Serbs, Hungarians, Czechs, Slovaks, Italians, and others. The most numerous were Orthodox Serb refugees who fled the Ottoman Empire in the late 17th century and settled in large enclaves in the Slavonian and Vojvodinan military border zones.

In 1689 and 1690, with the encouragement of Peć patriarch Arsenije III Črnojević, Orthodox Serbs in Ottoman Serbia and Kosovo staged rebellions in support of a Habsburg drive into their regions. (See Map 24.) When Ottoman forces expelled the invaders and reestablished control amid a welter of atrocities, some 40,000 Serbs, Patriarch Arsenije among them, fled across the Danube, ultimately settling in the newly established Habsburg Slavonia-Vojvodina military border region. Although later Serb nationalists claimed that the number of refugees involved approached 500,000, and that Arsenije heroically led them to safety, recent scholarship discredits those contentions.

The border Serbs of Vojvodina and Slavonia were able to pursue their native Orthodox culture within the Habsburg Western European state, while they also enjoyed exposure to the intellectual movements current in the West and, through their Orthodox church connections, in Russia. Despite periodic frictions with their Habsburg German and Hungarian hosts, the border Serbs developed an East-West European intellectual alloy that eventually spawned modern Serbian nationalism. The two cultural streams merged primarily within the ranks of the Serbian Orthodox church and spread to the lay believers, first in the border zones and then southward across the Danube among Serbs in the Ottoman Empire.

To ensure the numerous border Serbs' faithful service, Leopold granted them a variety of privileges, including landowning rights and assorted economic incentives. The most significant was autonomy for the Serbs' Orthodox church within the Habsburg Empire. In 1713 an independent Serbian archbishopric was established in Sremski Karlovci with jurisdiction over all Slavic-rite Orthodox believers in Vojvodina, Slavonia, and Pannonian Hungary. Its first primate was the refugee Peć patriarch Arsenije Črnojević.

The border Serbs' religious autonomy harmonized well with their former Ottoman *millet* traditions and reinforced their old *millet* sense of group identity. In the Orthodox world the church traditionally played a leading role in its followers' cultural, intellectual, and educational lives. Under the Sremski Karlovci archbishopric's patronage, schools, printing presses, and reading rooms were established in the local parishes and monasteries. Close cultural ties were forged with the Russian Orthodox church and, through it, the Russian government, whose growing imperialist interests in the Balkans made patronage of the Serbs potentially valuable. Throughout the 18th century the border Serb communities were inundated with Russian funds, books, and schoolteachers, and many Serb youths were granted scholarships for study in Russia. A lasting consequence of the Serbs' Russian connections during this period was their deeply ingrained Russophilism.

HABSBURG EMPIRE
HUNGARY
SLOVENIA
SLAVONIA
OTTOMAN EMPIRE
BOSNIA
DALMATIA
SERBIA
HERCEGOVINA
MONTENEGRO
KOSOVO
Sava R.
Drava R.
Danube R.
Bosna R.
Drina R.
Neretva R.
Ljubljana
Varaždin
Djurdjevac
Križevci
Zagreb
Virovitica
Szigetvár
Mohács
Senta
Rijeka
Karlovac
Sisak
Nova Gradiška
Brod
Vinkovci
Novi Sad
Petrovaradin
Sremski Karlovci
Zemun
Belgrade
Ogulin
Slunj
Senj
Otočac
Bihać
Banja Luca
Karlobag
Jajce
Travnik
Zadar
Šibenik
Split
Sarajevo
Srebrenica
Višegrad
Mostar
Novi Pazar
Dubřovnik
(Ottoman Vassal Client)
Kotor
Cetinje
Peć
Shkodër
MILES
0 10 20 30 40
0 20 40 60
KILOMETERS
Habsburg Croatia-Slavonia Border, late 18th Cent.
Habsburg Croatia-Slavonia Border, early 17th Century
Territories of the Habsburg Croatian-Slavonian Military Border
Territories of civil Croatia-Slavonia
Venetian Possessions

Map 24: The Ottoman Balkans, Late 17th–18th Centuries

Beginning in the second half of the 16th century, developments in Western Europe, with which the Ottomans were in constant contact, brought on progressive destabilization (Westerners term it "decline") within the Ottoman Empire.

New Western European technologies played havoc on the Ottomans' tradition-bound Islamic state. Beginning in the 16th century, Western naval developments ushered in the "Age of Discovery," which opened sea routes to the necessary and lucrative spice trade with Far East Asia, circumventing the Ottoman middlemen who formerly controlled such commerce. Gold and silver from the Americas flooded the empire's eastern Mediterranean markets, causing currency devaluation, rampant inflation, high taxes, and an explosive rush for cash at all levels of Ottoman society. Moreover, in the 17th century Western gunpowder technologies transformed the weaponry and tactics of warfare—a development to which the Ottomans, holding fast to traditional military approaches, were slow to respond. The result was the end of Ottoman military dominance in Europe and mounting defeats in battles with their European enemies. Western European states, such as France and Britain, were able to force changes in old commercial treaties—"capitulations"—on the Ottomans that placed nearly all of the empire's trade relations and profits in their own hands. Militarily antiquated and economically strangled, the Ottoman Empire ceased to expand, resulting in gradual, but perpetual, contraction in Europe.

A unique facet of Ottoman government in its ascendancy was that all of its bureaucratic and military offices were staffed by household slaves of the sultans, who held over them the power of life and death, making it the most centralized and efficient government in Europe from the 14th through the mid-16th centuries. It should be noted that slavery in the Ottoman context little resembled the degrading slavery of the American South. The sultan's slaves possessed and controlled immense power, wealth, social position, and public honor. Every government and military office was filled on the basis of individual merit, with no regard whatsoever for birth or social status. The system began breaking down in the mid-16th century, when pressures to increase state revenues led to the sale of offices to wealthy Muslim-born subjects, thus loosening the sultans' absolute authority over the government and military. The conditional military fiefs supporting the provincial forces, which technically were owned and distributed by the sultans, and on which the entire stability of the empire's military-administrative system depended, were sold or leased to the highest bidders without regard for military considerations. As the central government's control over the empire's institutions declined, bribery and corruption escalated, causing near military collapse and deteriorating living conditions for the Ottomans' Balkan subjects.

The empire's resulting territorial contraction in Europe began with the disastrous second Ottoman siege of Vienna (1683), following which the Habsburgs pushed the Ottomans out of Hungary and south of the Danube, except in the region of Banat, by 1699. The Ottomans also faced war with a new enemy—Russia—north of the Black Sea. The Habsburgs reopened military operations along the Danube in 1716, led by Prince Eugene of Savoy, and captured Belgrade (1717). In the Treaty of Požarevac (1718) the Ottomans lost Banat and relinquished part of northern Serbia and western Wallachia (Oltenia) to the Habsburgs. The Ottoman military, however, had not completely collapsed. In renewed warfare with the Habsburgs and Russia beginning in 1736, the Ottomans fought successful campaigns against both, winning back Belgrade, northern Serbia, and Oltenia in the Treaty of Belgrade (1739).

Wars among the Ottomans' European enemies granted them a respite from battle in the Balkans until 1768, when Catherine II the Great (1762-96) sent Russian troops into Moldavia and Wallachia. Sultan Mustafa III (1757-74) declared war but could not prevent the Russians from overrunning the two principalities or from conquering Crimea. Only the need to concentrate forces to crush the great Cossack revolt led by Pugachev (1773) diverted the Russian empress from her goal of conquering Istanbul and resurrecting an Orthodox *imperium* in the Balkans. The Treaty of Kyuchuk Kainardzha (1774), ending the hostilities, reflected the Ottomans' diminished status as a European Great Power. Russia gained extensive territories in the Black Sea region and free commercial navigation of its waters. A controversial clause also gave Russia representation at the Ottoman court on behalf of the empire's Orthodox subjects, while the Ottomans promised to protect Orthodox church property throughout the empire.

The combined effects of internal destabilization and successive military losses led to the breakdown of Ottoman central authority and to regional anarchy in the Balkans. There arose a group of strongmen *(ayans)* who carved out virtually independent realms for themselves and acted as autonomous regional rulers, jealous of their local independence but too weak to throw off entirely Ottoman suzerainty. By the late 18th century, ten *ayans* operated in the Balkans, two of whom proved particularly influential. Pasvanoğlu Osman Paşa controlled northwestern Bulgaria from Vidin and caused chaos in Serbia (1800), contributing to the 1804 Serbian uprising. Tepedenli Ali Paşa came to rule the single largest *ayan* Balkan province—Epiros, most of Macedonia and mainland Greece, Morea, the Ionian Islands, and southern Albania. He was so powerful that he dealt directly with the French, British, and Russians during the Napoleonic Wars and was a factor in igniting the Greek Revolution (1821). (See Map 25.)

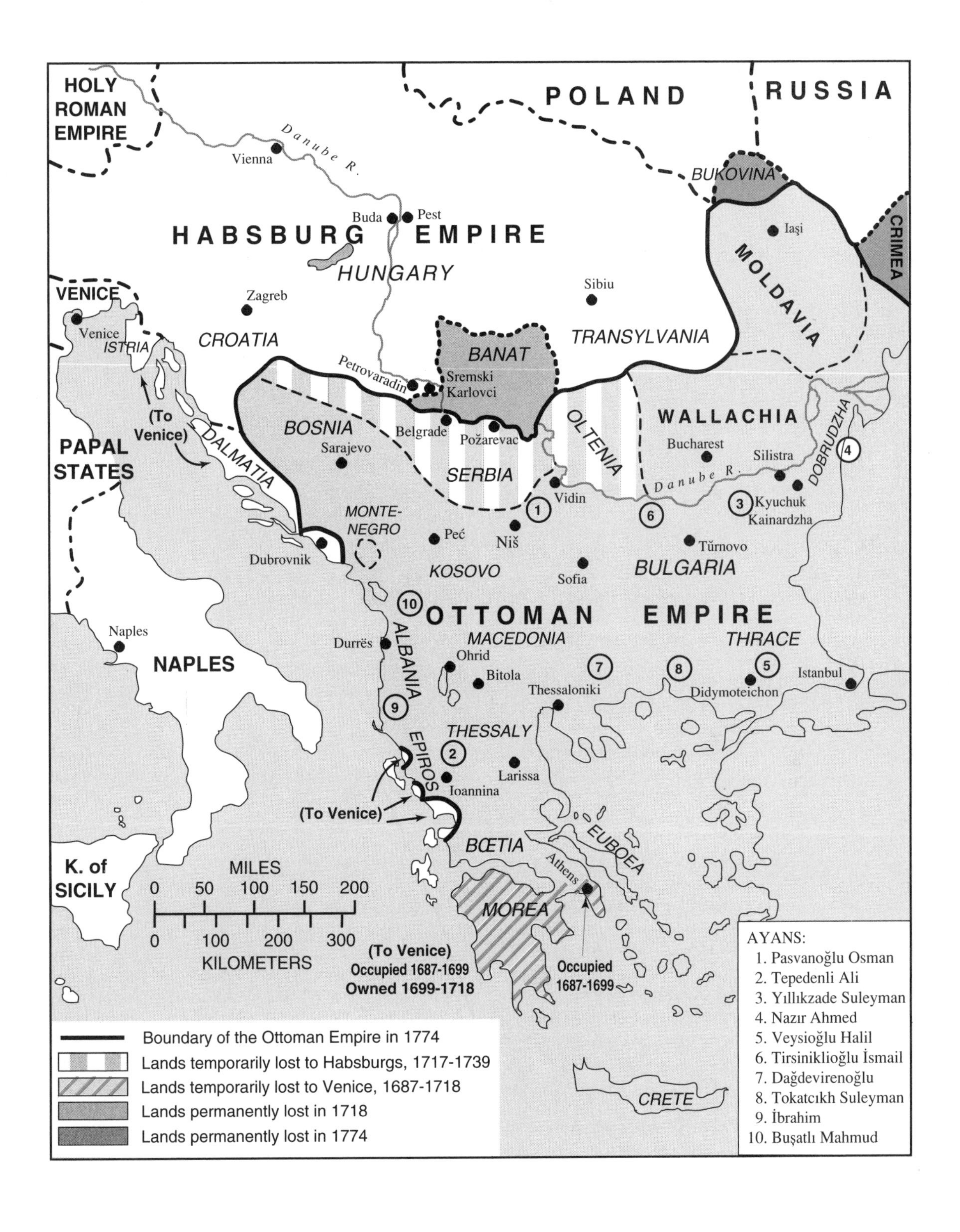

HOLY ROMAN EMPIRE
POLAND
RUSSIA
Danube R.
Vienna
BUKOVINA
Buda
Pest
HABSBURG EMPIRE
Iaşi
CRIMEA
MOLDAVIA
HUNGARY
VENICE
Zagreb
Sibiu
Venice
ISTRIA
CROATIA
TRANSYLVANIA
BANAT
Petrovaradin
Sremski Karlovci
(To Venice)
BOSNIA
Belgrade
Požarevac
OLTENIA
WALLACHIA
DOBRUDZHA
PAPAL STATES
DALMATIA
Sarajevo
Bucharest
Silistra
SERBIA
Danube R.
Vidin
Kyuchuk Kainardzha
MONTE-NEGRO
Peć
Niš
Tŭrnovo
Dubrovnik
KOSOVO
Sofia
BULGARIA
OTTOMAN EMPIRE
Naples
ALBANIA
MACEDONIA
THRACE
Durrës
Ohrid
NAPLES
Bitola
Thessaloniki
Didymoteichon
Istanbul
EPIROS
THESSALY
Larissa
Ioannina
(To Venice)
EUBOEA
BŒTIA
Athens
K. of SICILY
MILES
0 50 100 150 200
0 100 200 300
KILOMETERS
MOREA
(To Venice)
Occupied 1687-1699
Owned 1699-1718
Occupied 1687-1699
CRETE
Boundary of the Ottoman Empire in 1774
Lands temporarily lost to Habsburgs, 1717-1739
Lands temporarily lost to Venice, 1687-1718
Lands permanently lost in 1718
Lands permanently lost in 1774
AYANS:
1. Pasvanoğlu Osman
2. Tepedenli Ali
3. Yıllıkzade Suleyman
4. Nazır Ahmed
5. Veysioğlu Halil
6. Tirsiniklioğlu İsmail
7. Dağdevirenoğlu
8. Tokatcıkh Suleyman
9. İbrahim
10. Buşatlı Mahmud

Era of Romantic Nationalism
1804-1878

Map 25: Emergence of Modern Balkan States, 1804-1862

During the Napoleonic Wars, France won possession of Dalmatia, Istria, much of the Croatian Military Border, and most of Slovenia from the Habsburgs, organized them into the "Illyrian Provinces," and then annexed them (1809-13). French administration, law, and language were imposed; public works projects (road and bridge building, reforestation, land reclamation), and social reforms (serf emancipation, land redistribution) were initiated; and, most important, the tenets of liberal democracy and nationalism were put into actual practice. Although the French interlude lasted only a few short years, and its thrust essentially was anti-Croat and anti-Slovene (since the provinces were considered French and not Croatian or Slovenian), the Croats and Slovenes gained an awareness of their own ethnonational identities and briefly experienced firsthand the benefits of liberal, nationalist political culture.

In the Ottoman Balkans, nationalist concepts spread from Serb emigrants in Habsburg Slavonia-Vojvodina to the Ottomans' Serb subjects, transforming a local uprising begun in 1804 in the Belgrade region into a nationalist movement by 1815. The rebellion erupted in reaction to the anarchistic rule of local Belgrade Janissaries and originally was aided by the Ottoman central government. The Serb rebels were led by Djordje Petrović, a prosperous pig dealer with the nickname of Karadjordje (Black Djordje). The Habsburg military border Serbs sent volunteers and supplies to Karadjordje and attempted to establish a military alliance with Russia, which went to war with the Ottomans in 1806. The influence of the emigré Serbs soon transformed the rebellion from an attempt to reestablish legitimate Ottoman rule into a struggle for Serbian independence. When the Russians signed an armistice with the Ottomans in 1807, however, the rebels were left facing the full brunt of Ottoman forces. Reprieved by renewal of the Russo-Turkish war in 1809, the Serbs were left abandoned when Russia signed the Treaty of Bucharest (1812) so it could face Napoleon's impending invasion. The rebels then fell into internal disarray and the Ottomans reoccupied their territories in 1813. Karadjordje fled to the Habsburg Empire, and the uprising temporarily ended.

In 1815 the Serbs again rebelled, determined to establish complete independence from Ottoman control. The new rebel leader was Miloš Obrenović, a middle-class individual with an abiding personal hatred of Karadjordje, whom he accused of having poisoned his half brother. Through dogged military leadership, astute diplomacy, and highly refined bribery of Ottoman officials, Obrenović won the Ottomans' recognition as prince of an autonomous Ottoman province of Serbia in 1817. When Karadjordje returned from exile and voiced opposition to Obrenović's seemingly pro-Ottoman approach, he was assassinated by Obrenović's supporters, sparking a deadly blood feud between the two future Serbian royal families that plagued Serbian politics into the 20th century.

By virtue of their privileged commercial position in the Ottoman Empire, the Phanariote Greeks maintained direct relations with Western Europe and Russia. Greek trading colonies were founded throughout Europe and the Russian Black Sea coast and served as channels through which Western and Russian ideas spread to the Greek merchant class. With the founding of the *Philike hetairia* (Society of Friends) in 1814, secret revolutionary societies sprang up in Greek merchant colonies throughout Europe. They turned to Orthodox Russia for support of their plans, especially since one of their number, John Capodistrias, became Tsar Alexander I's (1801-25) foreign minister.

In 1821 a Greek nationalist revolution erupted when the Greek merchants' anti-Ottoman agitation coincided with Russian imperialist aims at dominating the Balkans and opening the Mediterranean to the Russian navy. A *hetairia* force, led by a Greek general in Russian service, Alexander Ypsilantis, unsuccessfully invaded the Romanian Principalities, which Greek nationalists mistakenly considered Hellenized by a century of Phanariote rule. Though a failure, news of the episode sparked a rebellion among the Greeks in the Peloponnese, who conducted an initially successful guerilla war against divided and weak Ottoman forces in the region.

The Ottomans' first reaction was to hang the Greek patriarch from the gate to his cathedral in Istanbul. Then, in 1824, they called in Egyptian troops, the only effective military forces left in the empire, to quell the uprising. In the midst of the fighting, the rebels fell to squabbling among themselves. As Egyptian successes mounted, Western Europeans, imbued with philhellenic sympathies and in the throes of the Romantic Movement, grew fed up with the slaughter of fellow Christians in Greece by Muslims. Western volunteers (including the poet Byron, who died at Mesalóngion) streamed into the region to join the rebels, whom they mistakenly considered the direct descendants of the classical ancients. Public opinion in Britain, France, and Russia eventually overcame their official predilection for nonintervention in revolutionary wars. Those states dispatched fleets to the eastern Mediterranean, and their combined intervention brought the destruction of the Egyptian-Ottoman fleet in Navarino Bay (1827), ultimately securing the Greeks' complete independence from the Ottomans in 1830.

By the Treaty of Edirne (Adrianople) in 1829, which ended yet another Russo-Turkish war (1828-29), Serbia was recognized as an autonomous Ottoman province governed by Obrenović. The Ottomans also recognized the independence of a small state of Greece at the London Conference (1830), governed from Athens by King Otto (Othon) I Wittelsbach (1832-62) from Bavaria. Wallachia and Moldavia became Russian protectorates until conditions following the Crimean War (1853-56) led to their unification as Romania. (See Map 37.)

BAVARIA
Danube R.
Vienna
RUSSIA
HABSBURG EMPIRE
Buda
Pest
BESSARABIA
Iaşi
MOLDAVIA
Ljubljana
SLOVENIA
HUNGARY
Zagreb
TRANSYLVANIA
Venice
Rijeka
CROATIA - SLAVONIA
ILLYRIAN PROVINCES
ROMANIA
(United 1862)
Galaţi
Zemun
VOJVODINA
BOSNIA
Belgrade
Sarajevo
DALMATIA
WALLACHIA
Bucharest
DOBRUDZHA
PAPAL STATES
SERBIA
Vidin
Danube R.
HERCEGOVINA
Niš
BULGARIA
Dubrovnik
MONTENEGRO
Cetinje
KOSOVO
Sofia
Plovdiv
OTTOMAN EMPIRE
Skopje
Edirne
Naples
ALBANIA
MACEDONIA
THRACE
Istanbul
Thessaloniki
KINGDOM OF THE TWO SICILIES
EPIROS
Ioannina
CORFU
THESSALY
IONIAN IS. REPUBLIC
(British Protectorate 1815-1864;
To Greece, 1864)
Mesalóngion
Thebes
Athens
Patras
Epidavros
Navplion
MILES
0 50 100 150 200
0 100 200 300
KILOMETERS
GREECE
Navarino
CYCLADES ISLANDS
CRETE
(To Egypt)
Border of Ottoman Empire
Autonomous Serbia, 1830
Independent Greece, 1830
Autonomous and united Romania, 1862
Napoleonic French Illyrian Provinces, 1809-1813
(returned to Habsburg control in 1814-1815)

Map 26: The Balkan Crisis of 1875-1876

The Ottomans were among the first non-European civilized societies to face direct competition from Western European scientific-technological, economic, and political-national development and it left them destabilized and weakened. Throughout the 19th century a number of sultans and a small group of perceptive government officials attempted to enact reforms intended to adapt to Western pressures while still retaining the basic premises of Islamic life. Their efforts collectively became known as the *Tanzimat*. Beginning with the suppression of the Janissaries (1826) by Mahmud II (1808-39), the *Tanzimat* became official policy in 1839 under Sultan Abdülmecid I (1839-61). Initially, adaptive reforms were aimed at "westernizing" the military and government administration and secularizing education (with only limited success). At the end of the Crimean War (1856), new measures were enacted to improve the lot of non-Muslim subjects to combat the spread of Western-style nationalism among them.

One post-1856 reform targeted *millet* reorganization, which had a significant impact on the Bulgarians. Abdülmecid ordered an Orthodox church council in Istanbul to reorganize the *millet* (1860-62). When it convened, Greek participants predominated and the Bulgarian clergy were underrepresented. Wealthy Bulgarian merchants in the capital then announced (1860) that Bulgarians no longer would recognize the Greek patriarch's authority and called for the creation of a separate Bulgarian Orthodox church, sparking the "Bulgarian Church Question." A bitter decade-long conflict between Bulgarians and Greeks began within the Orthodox *millet*. From the start, it was a nationalist dispute—no doctrinal positions were involved. The Bulgarians demanded their own church defining, in *millet* terms, the geographic extent of a Bulgarian ethnonational territory, while the Greeks viewed that demand as threatening Hellenism and the future of an enlarged Greece. Pressured by Russian ambassador Nicholas Ignatiev, in 1870 Sultan Abdülaziz (1861-76) recognized an independent Bulgarian church headed by an exarch seated in Istanbul—the Bulgarian Exarchate—with jurisdiction over large tracts of Bulgaria, Thrace, and Macedonia, and the ability to acquire further territories should two-thirds of their inhabitants vote to join. Given Bulgarian and Greek national aspirations, the Exarchate guaranteed future national conflicts over regions where the two sides' claims coincided. (See Map 30.)

The outbreak of an anti-Ottoman revolt in Bosnia and Hercegovina in 1875 provoked extremely tense European Great Power rivalries, known as the "Eastern Question." The conflicting interests of Britain and Russia lay at the "question's" heart. Russia's desire to end its essentially landlocked international situation by breaking through to the Mediterranean via the Bosphorus and Dardanelles straits was viewed by Britain as a threat to its maritime route to its most important Indian colony (by way of Suez). Russia sought to facilitate its policy either by directly controlling the eastern Balkans and the Straits or by gaining influence over any Orthodox Christian states in that region. Britain was determined to prevent such developments by upholding the Ottoman Empire as a buffer. A third Great Power protagonist—Habsburg Austria-Hungary—having been expelled from German affairs by Prussia in 1866, hoped to maintain its international position by expansion in the Balkans, so it also opposed any Russian presence in the region.

Centuries of the non-Muslim peasantry's exploitation by the landowning Muslim *beys* in Bosnia and Hercegovina led to the revolt. Habsburg and Russian officials in Dalmatia aided the rebels as best they could short of direct intervention, but when mediation efforts broke down, the fighting grew widespread. By early 1876 more than 150,000 refugees had flooded into surrounding states. The rebellion provided Serbia with an opportunity to shed Ottoman suzerainty entirely and commence territorial expansion. Serbia, in conjunction with Serb-inhabited Montenegro, declared war on the Ottomans in aid of the rebels, hoping to acquire the two rebellious provinces. Russia supported the Serbs' action, prompting Britain to intensify its efforts to preserve intact what remained of the Ottoman Empire in Europe. The Habsburgs sought to validate their Great Power status by dominating the Serbs and expanding their empire into Bosnia and Hercegovina. Vague reports of another anti-Ottoman uprising in Bulgaria (May 1876) added to the building international trepidations. (See Map 27.)

Tensions among the Great Powers increased in October when the Serbs unexpectedly were defeated by the newly "westernized" Ottoman forces. The Ottomans were prevented from occupying Serbia only by a Russian ultimatum demanding a cease-fire. In December Britain called a conference with Russia in Istanbul to force reforms on its Ottoman ally. British plenipotentiary Robert of Salisbury and Russian ambassador Ignatiev devised a compromise agreement that united Bosnia and Hercegovina into a single autonomous province and created two autonomous Bulgarian provinces. Montenegro retained its war gains in Hercegovina and Albania, and Serbia regained its prewar borders. Their negotiations, however, came to naught.

Amid public fanfare while the conference still was in session, new Sultan Abdülhamid II (1876-1909) proclaimed a liberal constitution, produced by the reform grand vizir Ahmed Şefik Midhat Paşa, rendering the plans from the conference ostensibly unnecessary. The Ottoman Constitution of 1876 was not simply a matter of expedient Ottoman duplicity (although it never truly was implemented). Although the Istanbul Conference forced the issue of its timing, the constitution was the creation of forces within Ottoman society that sincerely sought to broaden the *Tanzimat* reforms along highly westernized political lines.

RUSSIA
GERMANY
Danube R.
Vienna
AUSTRIA-HUNGARY
Budapest
BESSARABIA
Iaşi
Ljubljana
Venice
SLOVENIA
Zagreb
ISTRIA
TRANSYLVANIA
CROATIA-SLAVONIA
Galaţi
Zemun
BANAT
BOSNIA
ROMANIA
(Autonomous Ottoman Vassal)
Belgrade
DALMATIA
Sarajevo
SERBIA
(Autonomous Ottoman Vassal)
Bucharest
DOBRUDZHA
Mostar
HERCE-GOVINA
Vidin
Danube R.
Tŭrnovo
MONTE-NEGRO
Novi Pazar
Niš
Dubrovnik
Cetinje
Sofia
BULGARIA
ITALY
KOSOVO
OTTOMAN EMPIRE
To Austria-Hungary
Plovdiv
ALBANIA
Skopje
Naples
Edirne
Durrës
Ohrid
Istanbul
MACEDONIA
THRACE
Thessaloniki
EPIROS
Ioannina
THESSALY
Larissa
CORFU
Athens
SICILY
MILES
0 50 100 150 200
Patras
GREECE
0 100 200 300
KILOMETERS
CRETE
Border of territories under direct Ottoman control
Border of the Bulgarian Exarchate, 1872
Region of mass uprisings in Bosnia and Hercegovina, 1875-1876
Regions of Bulgarian "April" Uprising, 1876
Border of the Kingdom of Hungary within Austria-Hungary

Map 27: The "San Stefano" Balkans, March 1878

Britain and Russia knew that the 1876 Ottoman constitution was worthless, since Abdülhamid II had no interest in accepting liberal-democratic restraints and it was doubtful that the Ottomans actually could implement its terms. In any event, Russia incrementally moved toward war with the Ottoman Empire from the time that the Bosnian and Hercegovinian rebellion began. In July 1876 Russian and Habsburg foreign ministers met at Reichstadt to discuss the Balkan situation and reached an agreement that essentially divided the Balkans between them, with Austria-Hungary dominant in the western half of the peninsula and Russia in the east, should Serbia and Montenegro be defeated. While those talks took place, news of Ottoman massacres of Bulgarians in a separate uprising (May 1876) filtered into Britain and Russia from missionaries, journalists, and diplomatic agents.

After moderate Bulgarian nationalists emerged victorious in the "Bulgarian Church Question," a small group of emigré revolutionary nationalists operating out of Bucharest in Romania stirred up an uprising among Bulgarians within the Ottoman Empire, hoping to take advantage of the Ottomans' preoccupation with events in the northwest. The May rebellion was a sad, badly planned affair. The ill-armed and disorganized rebels did little more than publicly rally, sing patriotic songs, and butcher their mostly pacific Muslim neighbors. The Ottomans organized makeshift retaliation and crushed the rebellion. The brutality of their irregular *başıbazuks,* recruited from among local Muslims and Circassians inflamed by revenge for rebel atrocities against fellow Muslims, resulted in at least 15,000 Bulgarian deaths (later exaggerated to as high as 100,000) and widespread devastation.

Word of the *başıbazuks'* atrocities filtered to American-run Robert College, located north of Istanbul. Soon the Western diplomatic community in Istanbul was abuzz with rumors, which eventually found their way into newspapers in the West. Rumor-mongering news stories about Ottoman Muslim atrocities against Christians, however, were unwelcome in Britain, where Conservative Prime Minister Benjamin Disraeli's government was committed to supporting the Ottomans in an already tense Balkan situation. The Liberal opposition published slanted newspaper articles and pamphlets, attacking Disraeli's pro-Ottoman policy so effectively that Queen Victoria (1837-1901) publicly questioned the morality of Britain's pro-Ottoman stance. His hands tied, Disraeli informed the Ottomans that Britain could not defend them against possible Russian attack.

In January 1877 Russia brokered the Budapest Convention with Austria-Hungary, stipulating the latter's neutrality in case of hostilities with the Ottomans and implying the Habsburgs' acceptance of a Russian presence in the eastern Balkans in return for their right to occupy Bosnia-Hercegovina. With Habsburg neutrality secured and British resistance muzzled, the way was cleared for another Russo-Turkish war, which Tsar Alexander II (1855-81) declared in April 1877.

Romania, an Ottoman vassal state through which Russia's forces had to pass in invading the Balkans, eagerly signed an anti-Ottoman military convention with Russia, mobilized its forces, declared war on the Ottomans, and proclaimed its complete independence (May 1877).

Eschewing direct Romanian, Serbian, and Montenegrin military participation, Russian forces, joined by Bulgarian volunteers, invaded Ottoman Bulgaria in late June and penetrated southward to the Balkan Mountains. Ottoman resistance stiffened at the strategically located fortress-city of Pleven (Plevna) and a siege ensued. The humiliated Russians requested military assistance from Romania, Serbia, and Montenegro, but Pleven held out until December, after which the Russians resumed their advance and Ottoman resistance collapsed. As Russian forces pushed south in January 1878, a welter of atrocities was inflicted on the local Muslim population by the troops, Bulgarian volunteers, and emboldened local Bulgarian villagers. Some 260,000 Muslims perished and over 500,000 refugees fled with the retreating Ottoman forces.

At the end of February, Russian forces were in sight of Istanbul itself, and it seemed that Russia would finally realize its imperialist dream of acquiring the city and free access to the Mediterranean. Britain, whose public now understood the threat to British imperialist interests, reacted strongly, sending a fleet to the Straits with orders to intervene should the Russians attempt to snatch Istanbul. Taking the hint, Russia halted and forced the Ottomans to sign the Treaty of San Stefano (March 1878), granting Serbia, Montenegro, and Romania complete independence and creating a large Bulgarian state.

New Bulgaria included virtually all of the central Balkans. At the stroke of a pen, the Russians handed the Bulgarians possession of Bulgaria Proper (less Dobrudzha, which was given to Romania in compensation for Russia's annexation of Bessarabia), Thrace (less regions around Istanbul, Edirne, and Thessaloniki), and all of Macedonia, making Bulgaria the single largest Balkan state. By rewarding the Bulgarians' national aspirations with borders exceeding even those of the Exarchate, Russia expected a rich return. Unable to control the strategic Straits directly, Russia would leverage Bulgarian gratitude for access.

All of the European Great Powers and Balkan states raised a howl when San Stefano's terms became public. No Western European Power could accept a settlement handing Russia virtual control of the crucial Straits. Neither Serbia nor Montenegro was satisfied with its winnings, wanting all of the territories occupied during the war. Greece, kept out of the war by a British blockade, was incensed that Bulgaria was given Macedonian territories claimed by Greek nationalists. It was obvious that San Stefano's terms needed modifying to safeguard Europe's continued peace.

GERMANY
RUSSIA
Vienna
AUSTRIA - HUNGARY
Budapest
BESSARABIA
Iaşi
Cluj
Ljubljana
SLOVENIA
Zagreb
Venice
CROATIA-SLAVONIA
Zemun
Belgrade
BOSNIA
ROMANIA
(Independent)
Bucharest
DOBRUDZHA
Constanţa
Zadar
Sarajevo
SERBIA
(Independent)
Danube R.
Silistra
Vidin
Pleven
Varna
HERCE-
GOVINA
Novi
Pazar
Niš
BULGARIA
(Autonomous Ottoman Vassal)
MONTE-
NEGRO
Cetinje
Dubrovnik
Priština
KOSOVO
Sofia
ITALY
Peć
To Austria-
Hungary
Shkodër
Plovdiv
Skopje
Naples
ALBANIA
Edirne
Ohrid
Durrës
MACEDONIA
Kavala
San Stefano
Istanbul
Kastoria
Thessaloniki
OTTOMAN EMPIRE
EPIROS
Trikala
Larissa
Ioannina
CORFU
THESSALY
GREECE
SICILY
Athens
Patras
CYCLADES
IS.
MILES
0 50 100 150 200
0 100 200 300
KILOMETERS
CRETE
Territory given to autonomous Bulgaria
Territory lost by Romania to Russia
Territories gained by independent Montenegro, Serbia and Romania
Balkan territories remaining under direct Ottoman control

Map 28: The "Berlin" Balkans, July 1878

The Russians probably expected strident international opposition to the San Stefano Treaty. They were not disappointed. Britain particularly was belligerent. (The lyrics of a popular British song at the time—"We don't want to fight, but, by jingo, if we do, we've got the ships, we've got the men, we've got the money, too"—gave the English language a new term for bellicose nationalist foreign policy aggressiveness—jingoism.) Amid the widespread international uproar against the San Stefano Treaty, German chancellor Otto von Bismarck, eager to demonstrate young Germany's international weight, invited all concerned to a congress in Berlin to work out an acceptable final treaty.

Only the Great Powers' concerns were addressed at Berlin. Realizing that the Western Europeans present essentially were united in opposing Russian dominance in the eastern Balkans, Russia could only hope to retain as much of its gains as they would allow or risk a major European war. The interests of the Balkan states were listened to politely but had little weight in the deliberations. The Ottomans were reduced to spectators at their own funeral, ignored and insulted by all sides. On 13 July 1878, after intense negotiations among the participating Powers, the Ottomans and the small Balkan states received the final treaty's dictate.

Most of San Stefano's non-Bulgarian provisions were upheld at Berlin. All of the states that received their independence in San Stefano retained it (though Serbia and Montenegro actually lost territorial additions in the Berlin settlement); Russia retained its acquisition of Bessarabia; and Romania received its slice of Dobrudzha. Greece, which was not included under San Stefano, again received nothing at Berlin. (In fact, it lost territory, since it was forced to watch the Ottomans hand over control of Cyprus to Britain.) To the utter dismay of the Serbs and Montenegrins, and in a development completely unrelated in any direct way to San Stefano (but in accordance with the Reichstadt and Budapest conventions, as well as with a prevalent undercurrent of anti-Russian and anti-Orthodox sentiment among the Western Europeans), Habsburg Austria-Hungary was permitted to occupy rebellion-torn Bosnia-Hercegovina and the Sandjak of Novi Pazar, an Ottoman province that separated Serbia from Montenegro (thus preventing their possible future unification into a single "Great Serbian" state).

It was the Bulgarian provisions of San Stefano, which had necessitated the Berlin Congress in the first place, that underwent drastic modification. The large principality created by the Russians was dismantled to deprive Russia of its strategic advantages relative to the Straits and the eastern Balkans. San Stefano's "Great Bulgaria" was sliced into four parts. Bulgaria Proper, mostly lying north of the Balkan Mountains, emerged as an autonomous Bulgarian Principality ruled by an elected prince who governed under technical Ottoman suzerainty. The region south and southeast of Bulgaria and northwest of Ottoman Thrace was designated a new autonomous Ottoman province of Eastern Rumelia, with its capital located in Plovdiv. Rumelia's governor was to be a Christian appointed by the sultan and approved by the Great Powers. Western Thrace, which lay along the shores of the Aegean south of Eastern Rumelia, was returned to direct Ottoman control, thus cutting off any possible Bulgarian access to that sea. Also returned to direct Ottoman administration was the entire region of Macedonia.

While the Berlin settlement may have headed off the immediate war crisis by satisfying the imperialist concerns of the Western European Great Powers over Russia's successes in the Balkans, it created heated and deep-seated dissatisfaction among all of the small Balkan states. The dismemberment of San Stefano's "Great Bulgaria" struck the Bulgarians' short-lived national jubilation like a hammer blow. Euphoria swiftly changed to disillusionment and then to stubborn resolution to win back that which had been lost at Berlin. Their faith in the Orthodox Russians was shaken, even though the new liberal-democratic Bulgarian government was shaped with Russian encouragement, Russians held most of the prominent positions in the principality's ministries, and the infant Bulgarian military was trained and officered by Russians. The strong Russian presence in post-Berlin Bulgaria was intended to dampen the growing Bulgarian national resentment of the Berlin settlement but, in the end, would prove incapable.

The Serbs, forced to accept Habsburg occupation of territories they considered their national preserve, felt deeply betrayed by their longtime Russian allies, who had shown themselves willing to give away to Bulgarians Macedonian regions that were just as important to Serbian national aspirations as Bosnia and Hercegovina. The Serbs found it expedient to make an accommodation with the Habsburgs to free their hands for dealing with their new Bulgarian national rivals in the south. The Greeks, their national territorial ambitions having been denied at Berlin, were resolved to make every effort to win what they considered their rightful borders in the north.

The nationalist ambitions of all Balkan peoples would collide violently in the decades after the Berlin Treaty, the Western Europe-imposed terms of which served as the fundamental driving force in the peninsula's subsequent divisive events. (Those same Western Europeans later came to label the divisions disparagingly as "balkanization.") Nor would Europe in general be spared the consequences of the radicalized Balkan nationalisms generated by the Berlin settlement. By handing Bosnia-Hercegovina and the Sandjak of Novi Pazar over to Austro-Hungarian control, thus stymieing Serbian national aspirations, the Berlin Congress proved to be the first step down the road to the global horrors of World War I.

RUSSIA
GERMANY
Danube R.
Vienna
AUSTRIA - HUNGARY
Budapest
BESSARABIA
Ljubljana
Cluj
Iași
Zagreb
Venice
ISTRIA
CROATIA-SLAVONIA
TRANSYLVANIA
Zemun
BOSNIA-HERCEGOVINA
BANAT
ROMANIA
(Independent)
DOBRUDZHA
Constanța
Belgrade
SERBIA
(Independent)
Bucharest
DALMATIA
Split
Sarajevo
SANDJAK
Novi Pazar
Niš
Vidin
Danube R.
Silistra
Varna
BULGARIA
(Autonomous Vassal)
Tŭrnovo
MONTE-NEGRO
Cetinje
Dubrovnik
Peć
Priština
KOSOVO
Sofia
EASTERN RUMELIA
(Autonomous)
ITALY
To Austria-Hungary
Shkodër
Skopje
Plovdiv
OTTOMAN EMPIRE
Naples
Durrës
ALBANIA
Ohrid
Edirne
MACEDONIA
Kavala
THRACE
Istanbul
Kastoria
Thessaloniki
EPIROS
Ioannina
Larissa
CORFU
THESSALY
GREECE
SICILY
Athens
Patras
CYCLADES IS.
MILES
0 50 100 150 200
0 100 200 300
KILOMETERS
CRETE
Border of Bulgarian autonomous Ottoman vassal state
Border of autonomous Ottoman province of Eastern Rumelia
San Stefano territories lost at Berlin by Montenegro, Serbia, and Bulgaria
Ottoman territories placed under Austro-Hungarian occupation

Era of Nation-State Nationalism
1878-1944

Map 29: Balkan State Territorial Expansion, 1881-1886

All of the post-Berlin Balkan nation-states shared a common national imperative: To satisfy their "right ful" national territorial ambitions within the context of existing Great Power relationships. Although Berlin was convened to determine those relationships, discussions were couched almost exclusively in terms of Balkan national development. By officially recognizing the national independence of the four existing states, by sanctioning the creation of a fifth (autonomous Bulgaria) at the Ottoman Empire's expense, and by giving highest priority to matters of those states' territories and borders, the Great Powers unwittingly signaled that the Balkans' future lay with the Ottomans' Christian European inhabitants.

Serbian prince Milan Obrenović (1868-89) correctly recognized Berlin as a major diplomatic defeat for Russia that crippled its Great Power standing. He saw no alternative for Serbia but to establish closer ties to Austria-Hungary, despite the Habsburg occupation of Bosnia-Hercegovina. Milan brought Serbia into the Habsburg orbit through a series of railroad, commercial, and political agreements, all of which transformed Serbia into a dependent Habsburg satellite.

In Romania, Russia's acquisition of Bessarabia at Berlin rankled, especially in light of the Romanian military assistance given Russia during the recent war under treaty terms guaranteeing Romanian territorial integrity. Prince Carol (Karl) I of Hohenzollern-Sigmaringen (1866-1914), a German, turned to Austria-Hungary for support against a perceived Russian threat to Romanian sovereignty. Despite the deterioration in the Habsburg Transylvanian Romanians' position following 1867 and Habsburg economic threats concerning control of Danube River traffic, Carol signed a secret anti-Russian defensive military treaty with Austria-Hungary (1883), which associated Romania officially with the Central Alliance (Germany, Austria-Hungary, and Italy) against the Entente Alliance (Russia and France).

By the Berlin Treaty's terms, the new Bulgarian Principality existed under loose Ottoman suzerainty but was ruled directly by an elected prince governing through a constitution and a representative assembly. In early 1879 an assembly of nationalist delegates from various Bulgarian-inhabited regions met in Tŭrnovo with Russian encouragement to devise the new state's political system. Heated debates produced a highly liberal-democratic constitution and a mandate to reacquire all San Stefano lands lost at Berlin. Sofia was designated Bulgaria's capital because of its proximity to Macedonia and its position as a major communications crossroad. As first Bulgarian prince, the assembly chose young Alexander (Aleksandŭr) I of Battenberg (1879-86), a German acceptable to all of the Great Powers. Jealous of sharing power with an elected assembly, Aleksandŭr attempted to cement his position within the state by pursuing Bulgarian nationalist territorial expansion. The opportunity arose in the new Ottoman province of Eastern Rumelia.

Eastern Rumelia was inhabited predominantly by ethnic Bulgarians, who succeeded in gaining control of most crucial provincial offices. In 1885 the Bulgarian-dominated Rumelian militia overthrew the province's government and proclaimed unity with Bulgaria. Challenged by the fiery former revolutionary and ardent nationalist speaker of the national assembly, Stefan Stambolov, to go either to Plovdiv (Rumelia's capital) or back to Germany, an initially hesitant Aleksandŭr ordered the Bulgarian army into the province and journeyed to Plovdiv, where he embraced the union.

Unification was accomplished over the vehement protests of Russia, which withdrew all of its advisors, ministers, and military officers from Bulgaria. In late 1885 Russia's actions prompted Serbia, which feared that nationalist momentum from the unification would carry the Bulgarians into Macedonia, to declare war on Bulgaria. The Serbs expected an easy victory and territorial acquisitions. Instead, the Bulgarians repulsed the Serbian invasion forces and then assaulted Serbia in turn. Only Habsburg threats to intervene on the Serbs' behalf stopped the Bulgarian invasion, and a peace treaty was signed in Bucharest on 3 March 1886 (the eighth anniversary of San Stefano), securing the unification of Rumelia with Bulgaria.

Greece had been ignored by the Great Powers at Berlin. In return for the Ottomans being "invited" to turn parts of Thessaly and Epiros over to Greece, the Greeks had been forced to accept Britain's occupation of Cyprus. Such perfunctory treatment demonstrated that Greece was considered of little account in the predominantly Habsburg-Russian contention for Balkan hegemony that emerged at Berlin. Because of its location at the Balkans' extreme southern tip, Greece fell more within Britain's sphere of interests, including eastern Mediterranean maritime routes. Britain, however, considered its commitment to the Ottomans of more immediate importance than Greece.

In 1881 the Great Powers convinced the Ottomans to accept their Berlin "invitation" and grant Greece nearly the whole of Thessaly. Despite British efforts, most of Epiros was not included in the territorial transfer because of Great Power interest in an emerging Albanian national movement in the region. When Greece mobilized its military forces (1886) to invade Ottoman Epiros and take it as "compensation" for Bulgaria's unification, the Great Powers, led by Britain, blockaded the state until Greece's troops were stood down. Later (1897), Greece, though militarily unprepared, declared war on the Ottomans in support of an ethnic Greek uprising (1896-97) on Ottoman Crete. Great Power objections to the war prevented any other Balkan state from joining Greece, which then suffered an ignominious defeat. Last-minute Great Power intervention spared the Greeks the full measure of Ottoman retaliation.

RUSSIA
GERMANY
Danube R.
Vienna
AUSTRIA - HUNGARY
Budapest
BESSARABIA
Cluj
Iaşi
Ljubljana
SLOVENIA
Zagreb
Venice
ISTRIA
CROATIA-SLAVONIA
TRANSYLVANIA
Zemun
BANAT
BOSNIA-HERCEGOVINA
Belgrade
ROMANIA
DOBRUDZHA
Bucharest
Constanţa
DALMATIA
SERBIA
Sarajevo
Split
SANDJAK
Novi Pazar
Danube R.
Silistra
BULGARIA
MONTE-NEGRO
Niš
Tŭrnovo
Dubrovnik
Sofia
Cetinje
Priština
EASTERN RUMELIA
ITALY
KOSOVO
Plovdiv
Shkodër
To Austria-Hungary
Skopje
Naples
OTTOMAN
Edirne
Durrës
Ohrid
ALBANIA
MACEDONIA
THRACE
Istanbul
Thessaloniki
Bitola
EMPIRE
Larissa
Ioaninna
THESSALY
GREECE
SICILY
Athens
Patras
MILES
0 50 100 150 200
0 100 200 300
KILOMETERS
CRETE
United with Bulgaria, 1885; union recognized 1886
Acquired by Greece, 1881
Occupied by Austria–Hungary

Map 30: The Macedonian Question

Between the Congress of Berlin (1878) and the Balkan Wars (1912-13), nationalist political affairs in the Balkans were dominated by the "Macedonian Question." It was a conflict among Bulgaria, Greece, and Serbia—three young states whose territorial aspirations had been disappointed or ignored by the European Great Powers meeting in Berlin—for possession of the Ottoman province of Macedonia, a region slightly larger than Vermont. At Berlin, the Russian-dictated San Stefano borders of Bulgaria, which encompassed Macedonia, were overthrown, and the state was reduced to a fraction of the size Bulgarians considered acceptable. The Serbs had been forced to relinquish some territories won in the 1877-78 Russo-Turkish War and were compelled to accept Austro-Hungarian occupation of Bosnia-Hercegovina, a region they adamantly claimed as Serb national territory. (See Map 28.) The Greeks, who had been restrained from participating in the recent war against the Ottomans, felt insulted by Britain's occupation of Cyprus, which they claimed as their own. All three turned toward Macedonia as compensation for their perceived losses at Berlin.

The Bulgarians advanced strong arguments supporting their claim. Macedonia had been an integral part of the first Bulgarian state (681-1018), during which its regional capital of Ohrid became a leading Slavic cultural center and seat of the first independent Slavic Orthodox church, the Bulgarian Archbishopric-Patriarchate of Ohrid. Later, under Tsar Samuil, Macedonia constituted the core of the Bulgarian state. (See Maps 8 and 9.) The language of the region's Slavs was so similar to Bulgarian that it was considered a dialect rather than a separate tongue. Some leading exponents of the 19th-century Bulgarian national revival, such as the Miladinov brothers, were from Macedonia. The national movement evolved into the "Bulgarian Church Question," which succeeded in winning Ottoman recognition of a Bulgarian *millet* through institution of an autonomous Bulgarian church—the Exarchate—separate from the Orthodox *millet* controlled by the Greek Patriarchate of Constantinople. (See Map 26.) Since the Ottomans decreed that any region where two-thirds of the population voted to join would fall under the authority of the new Bulgarian church, most Macedonian Slavs eventually voted to join the Exarchate, thus removing themselves from direct Greek ecclesiastical control.

Greek claims to Macedonia were grounded in allusions to ancient Macedon and its two famous rulers, Philip II (359-36 B.C.E.) and Alexander III the Great (336-23 B.C.E.), but primarily were focused on Byzantine possession of the region and later Greek control of the Ottoman Orthodox *millet*. (See Map 22.) Following independence from the Ottomans, nationalist Greeks concocted the "Great Idea" *(Megale idaia)*, a political program calling for restoration of the Byzantine Empire as the natural Greek nation-state. The borders would be defined by territories in which the Greek language dominated within the Ottoman Orthodox *millet*. After the Ottomans disbanded the Bulgarian Archbishopric-Patriarchate of Ohrid (1767), Macedonia lay under direct Greek patriarchal control and a modicum of Hellenization occurred. To Greek nationalists, Macedonia rightfully was Greek. During the Bulgarians' campaign to convince the Macedonians to join the Exarchate, the Greeks countered with efforts of their own. The situation inside of Macedonia quickly degenerated into Bulgarian-Greek violence and terrorism, with native Macedonians becoming the primary victims.

Barred from expansion into Bosnia-Hercegovina, the Serbs were forced to look to Macedonia for possible future expansion. They too possessed historical claims. During the reign of Serbian Car Stefan Dušan, his capital was the Macedonian city of Skopje and Macedonia formed the heart of his state's lands. Dušan's empire served as the territorial model for the modern state desired by the Serbian nationalists. (See Map 17.) Failing to win lands from the Bulgarians by force in 1885-86, the Serbs felt compelled to enter the fray for Macedonia's possession. Serbian bands joined those of the Bulgarians and Greeks in the ethnic fighting that plagued and terrorized Macedonia's natives.

In 1893 Macedonian Slavs formed a revolutionary-nationalist organization of their own—the Internal Macedonian Revolutionary Organization (IMRO)—to oppose both the violence of the outsiders' nationalist bands and continued Ottoman control. Its members mostly were native Macedonians with some Bulgarians. IMRO's program was "Macedonia for the Macedonians," and, despite periodic dominance within its leadership by pro-Bulgarian elements within the organization, it signaled the emergence of a new, strictly Macedonian nationalist movement. IMRO's nationalists essentially coopted the Bulgarian historical argument but substituted "Macedonian" for "Bulgarian" when referring to ethnicity. In the 1890s a second Macedonian group—the External Macedonian Revolutionary Organization (EMRO)—emerged as IMRO's nationalist rival, but EMRO's blatantly pro-Bulgarian program (which enjoyed the backing of Bulgaria's government and military) and its use of indiscriminate violence won little support within Macedonia.

Violent clashes among nationalist parties in Macedonia became endemic, while attacks on Ottoman authorities multiplied. In 1903 IMRO sparked the unsuccessful anti-Ottoman Ilinden Uprising, which resulted in a futile intervention by the Western Great Powers to stop the bloodshed. Constant violence and terrorism uprooted thousands of Macedonian Slavs, most of whom fled to southwestern Bulgaria, where they established a virtual state-within-a-state and became a militant force in Bulgarian politics. By 1912 all of the protagonists in the Macedonian struggle came to realize that the Ottoman presence had to be eliminated before any further nationalist solution could be achieved. With Russia's urging, Bulgaria, Serbia, and Greece put aside their antagonisms long enough to form an anti-Ottoman alliance aimed at expelling the Ottomans from Europe and settling the Macedonian problem. The results were the Balkan Wars. (See Map 33.)

Current international borders

Historic Macedonian boundary (Bulgarian claim)

Approximate Serbian claims

Approximate area claimed by both Greece and Serbia

Approximate Greek claims

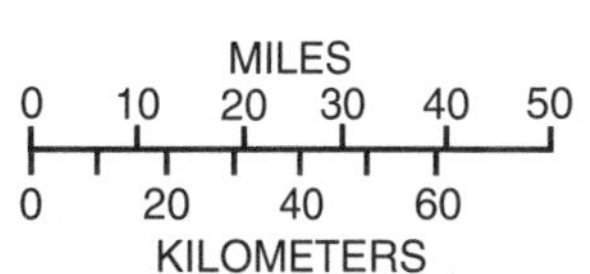

Map 31: The Balkans, 1908

A Turkish nationalist movement emerged in the Ottoman Empire in the 1860s when westernized Turks, critical of traditional governing policies and opposed to superficial reforms, were forced to flee abroad. Establishing themselves in Paris, by 1902 they acquired the name of "Young Turks" and divided into two factions. One stood for centralizing the existing Ottoman state under Turkish dominance, and the other called for decentralization and full ethnic autonomy for all of the empire's subjects.

While the Young Turks argued in Paris, officers of Turkish units stationed in Macedonia acted. The movement, grounded in an empire-wide secret military officers' organization called the Society of Liberty, headquartered in Thessaloniki, espoused the extremist Young Turks' ultranationalist centralization program. Well organized and supported by their troops, the Macedonian officers, led by Enver Paşa and including Mustafa Kemal, revolted in 1908 to forestall a Western Great Powers plan to intervene in Macedonia and halt the continued unrest there. Fearing a possible partition of the empire, the military leaders initiated a well-considered plan. An ultimatum to implement Midhat Paşa's 1876 reform constitution, which never had been put into force, was telegraphed to Sultan Abdülhamid II. (See Map 26.) The military units in Macedonia demonstrated their support and the sultan was constrained to comply.

The jubilation of Westerners and non-Muslims in the empire turned to anger and fear as it became apparent that the Young Turks were intent on preserving the old empire as a Turkish nation-state. The revolutionaries, organized as the Committee for Union and Progress (CUP), subordinated the sultan to their will. They initiated a policy of centralization and Turkish hegemony formerly unknown in the Ottoman Empire and counter to the spirit of the constitution that they ostensibly had risen to instate. Virtually every non-Turkish subject population was forced to react against the new regime, spawning the nationalist awakenings of the Albanians, Arabs, and Armenians. The Young Turks' pseudo-Western ultranationalist program ultimately led to atrocious massacres among their subject populations (such as the Armenians) and the swift disintegration of the Ottoman Empire during and after World War I. (See Maps 34 and 35.)

The year 1908 also witnessed the outright annexation of Bosnia-Hercegovina by Austria-Hungary. It had implemented its Berlin right to occupy Bosnia-Hercegovina almost immediately after the treaty (1878). (See Map 28.) The occupiers initiated a civic works regime not seen in the Balkans since the French Illyrian episode. By all Western standards, Habsburg occupation should have been a boon for the inhabitants. Unfortunately, such was not the case. The Habsburgs mistakenly failed to dismantle the old Muslim-controlled landholding regime that had evolved there over the centuries. A few thousand Muslim *beys* continued on as large estate owners and wielded immense local power over tens of thousands of Christian peasants. The estates were run in obsolete fashion, with inefficient methods of land use and outmoded equipment and techniques. Agriculture, the region's primary economy, remained backward and the population poor. Despite all the other benefits of Austrian occupation, the majority of the population—Orthodox and Roman Catholic Christians—remained downtrodden and grew increasingly discontented.

The 1908 Young Turk Revolution threw Europe into turmoil, and the Great Powers hastily convened foreign policy meetings among themselves in attempts to reshuffle the balance of power in the Balkans should the Ottoman state completely collapse. In September a meeting between the foreign ministers of Austria-Hungary and Russia took place in Buchlau, Austria, at which the Austro-Hungarian foreign minister Count Alois Aehrenthal finessed his Russian counterpart Alexander Izvolsky into accepting the Habsburgs' outright annexation of Bosnia-Hercegovina in exchange for empty words regarding future Austrian support for Russia's claim to free access to the Mediterranean—essentially nothing. In October 1908 Austria-Hungary announced its intention to annex Bosnia-Hercegovina permanently.

The announcement was met by the Great Powers' strongly voiced concerns and by frenzy and rage among the Serbs. The Russians protested that they had been duped. Germany supported its Habsburg ally, while France and Britain stood by Russia. None acted, however, since all feared that a general European war would result (given the Central Powers-Entente alliance system then in place). In 1909 the CUP Ottoman government accepted the annexation in return for compensation from Austria-Hungary, and the crisis ended.

Taking advantage of the Great Powers' preoccupation with the annexation crisis, Bulgarian Prince Ferdinand I (1887-1918) declared Bulgaria completely independent of Ottoman suzerainty, which freed him to pursue ambitions in Macedonia. A bloody military coup in Belgrade in 1903 had overthrown the pro-Habsburg Serbian Obrenović dynasty by butchering King Aleksandr (1889-1903) and installing the Russian-looking Petr I Karadjordjević (1903-21), but Russia's problems with Japan forced the new Serbian ruler to attempt an accommodation with Bulgaria to resist mounting Habsburg hostility. The nationalist and anti-Habsburg policies of Nikola Pašić, Petr's prime minister, provoked a tariff conflict with Austria-Hungary in 1906 known as the "Pig War" (since Serbia's primary exports were pork products, and Austria-Hungary was the chief market). The rapprochement between Serbia and Bulgaria was short-lived, floundering over the issue of Macedonia, and seemed doomed once Ferdinand made his bid for independence.

GERMANY
Danube R.
Vienna
GALICIA
RUSSIA
AUSTRIA - HUNGARY
Budapest
Iaşi
BESSARABIA
Ljubljana
Cluj
SLOVENIA
Zagreb
Venice
ISTRIA
CROATIA-SLAVONIA
TRANSYLVANIA
Novi Sad
BOSNIA - HERCEGOVINA
BANAT
ROMANIA
Belgrade
DOBRUDZHA
Sarajevo
Bucharest
DALMATIA
SERBIA
Constanţa
Split
SANDJAK
Danube R.
Novi Pazar
Niš
Varna
MONTENEGRO
Tŭrnovo
Dubrovnik
KOSOVO
BULGARIA
Cetinje
Prizren
ITALY
Skopje
Sofia
Shkodër
Plovdiv
To Austria-Hungary
Naples
OTTOMAN
Edirne
ALBANIA
Durrës
Ohrid
MACEDONIA
THRACE
Istanbul
Thessaloniki
EPIROS
Ioannina
EMPIRE
Larissa
CORFU
SICILY
Athens
Patras
GREECE
MILES
0 50 100 150 200
0 100 200 300
KILOMETERS
CRETE
Bosnia-Hercegovina annexed by Austria-Hungary
Sandjak of Novi Pazar occupied by Austria-Hungary
Independent Bulgaria

Map 32: Bosnia-Hercegovina, 1908-1914

In 1878, when Bosnia-Hercegovina was occupied, its population was divided among three component elements. The single largest group was the East European Orthodox (some 43 percent), followed by the Muslim (39 percent), and finally the Western European Catholic (18 percent). Both Christian components had developed ethnonational self-identities through influences that had infiltrated into Bosnia from its neighbors—the Orthodox espoused a Serb identity and the Catholics a Croat one. Given Islam's traditional theocratic culture, the Muslims, though ethnically Slavic and speaking the same language as the Christians, held no ethnonational affiliation. They maintained an Islamic cultural self-identity alone.

In both Croatia and Bosnia-Hercegovina, Croatian and Serbian nationalists' reaction to the 1908 annexation was strong. The more moderate Croat "Yugoslav" (South Slav) nationalists saw it as opening a bright future, in which their dreams for creating an autonomous Croat-led Yugoslav state within the Habsburg Empire would be fulfilled. The heir to the Habsburg throne Archduke Francis Ferdinand had made it known that he favored the idea of restructuring Austria-Hungary once he attained power by extending to Czechs and possibly to Croats political autonomy similar to that enjoyed by the Hungarians. Radical Croatian nationalists considered the annexation the first step toward a "Greater Croatia," but for similar reasons a recently founded Serbo-Croat coalition considered it a catastrophe. To them, Francis Ferdinand's sympathies for a tripartite state system marked him as the personification of the Austro-Hungarian threat to their aspirations for complete separation from the Habsburgs. Nurtured by Serbia, the Serbo-Croats bewailed the annexation as an insult to "Yugoslavism" (by which they meant, whether the Croat partners realized it or not, Serbian nationalism). Repression by Hungarian administrators in Croatia only discredited the Austro-Hungarian administration in the eyes of most Croats and Serbs and intensified the Belgrade-looking faction's discontent.

By 1909 a new element threatening Austro-Hungarian rule in Bosnia-Hercegovina began to make itself felt among the Serb population. In 1902 a cultural society called *Posveta* (Enlightenment), funded in part by money from Serbia, was established for the express purpose of educating peasant and lower-class Serb children. Within a decade of its founding, the society spawned a new type of Bosnian Serb intellectual—poor, often jobless, with no vested interest in the Austrian-imposed establishment (the Habsburg administration usually hired Croats over Serbs), and resentful of the inequitable existing social system.

This younger generation of Bosnian Serbs formed the cadres of a movement known as "Young Bosnia," an amorphous but widespread association that sought independence from the Habsburgs and social reforms within a Bosnian Serb nation-state. There was little agreement among its members over strategy, but a general affinity among them for Russian revolutionary literature led them to follow mostly Russian models. Turning their backs on political reform tactics, the Young Bosnians embraced terrorism, which they elevated into a veritable cult. Terrorist acts and the "martyrs" that such acts invariably created inflamed their blood and inspired their efforts. By 1912 members of Young Bosnia were in direct contact with a secret Serbian ultranationalist revolutionary organization, commonly known as the "Black Hand" but officially named "Union or Death" *(Ujedinjenje ili Smrt).* The Black Hand was controlled by military officers holding high positions in the Serbian government—the very ones who had conspired to kill King Aleksandr Obrenović in 1903—and led by Colonel Dragutin Dimitrijević (also known as "Apis"). While King Petr disliked the Black Hand leadership personally (they were, after all, regicides) and they, in turn, operated beyond his control and often at variance with his policies, he tolerated the organization's existence because of its anti-Habsburg and "Greater Serbia" stance. Through its Young Bosnia contacts, the Black Hand engineered and armed terrorist activities inside of Bosnia.

In the five years preceding the outbreak of World War I in 1914, the Habsburg authorities in Bosnia came down hard on Young Bosnian agitation, making hundreds of arrests for treason and espionage and mostly winning convictions. In response, the youthful Bosnian Serb revolutionaries intensified their violence. When, in the early summer of 1914, it was announced that Archduke Francis Ferdinand would tour Bosnia and would visit Sarajevo on 28 June, *Vidovdan* (St. Vitus's Day)—the anniversary of the Battle of Kosovo Polje (1389), which Serbian nationalists considered morbidly sacred ("Remember our defeat so that we will never let it happen again")—neither the Young Bosnians nor the Black Hand could let pass the opportunity to murder that personification of the Habsburg threat to "Greater Serbia" national aspirations. When Francis Ferdinand arrived in Bosnia, the Young Bosnian, Gavril Princip, managed to shoot the Habsburg heir to the throne (and his wife). Princip's handgun shots on a Sarajevo street corner proved to be but the first of a thunderous barrage and the two bodies of his victims the first of millions of corpses that spanned four years, as the Central Powers-Entente alliance system refused to permit Austria-Hungary to punish Serbia for the Sarajevo crime by means of a limited, localized military drubbing. Within a month of the outbreak of the Habsburgs' war of retribution, begun in July 1914, the struggle mushroomed into all-out world war. (See Map 34.)

AUSTRIA
HUNGARY
Varaždin
Ljubljana
Zagreb
SLOVENIA
Drava R.
Danube R.
VOJVODINA
Osijek
Rijeka
SLAVONIA
CROATIA
Vukovar
Novi Sad
Sava R.
Banja Luka
Bihać
Belgrade
BOSNIA
Tuzla
Bosna R.
Drina R.
Jajce
Travnik
Zenica
Srebrenica
Zadar
SERBIA
Sarajevo
DALMATIA
Šibenik
Višegrad
Split
Neretva R.
SANDJAK
Mostar
Drina R.
Novi Pazar
HERCE-GOVINA
MILES
0 10 20 30 40
0 20 40 60
KILOMETERS
MONTENEGRO
KOSOVO
Dubrovnik
Kotor
Cetinje
Bosnia-Hercegovina
Border between Austria and Hungary
Occupied by Austria-Hungary until 1912
Sandjak and Kosovo were part of the Ottoman Empire
ALBANIA

Map 33: The Balkan Wars, 1912-1913

The Young Turks' repressive Turkish nationalist policies played into the hands of the Balkan state nationalists, eventually permitting Bulgaria, Serbia, Greece, and Montenegro to overcome briefly their mutual animosities and form an anti-Turkish military alliance in 1912 with Russian encouragement. Taking advantage of the Turks' involvement in a war with Italy over Tripoli (1911), Serbia and Bulgaria hammered out a military treaty of mutual assistance in early 1912. A secret annex dealt with the future fate of Balkan regions still under Turkish control. Serbia was to receive the Sandjak of Novi Pazar, Kosovo, and a large strip of northern Macedonia. Western Thrace was ceded to Bulgaria. The bulk of Macedonia was to form an autonomous province. Should the autonomous province prove unworkable, the secret annex provided for its further division, with Bulgaria and Serbia each receiving additional territories and the remaining areas subject to Russian arbitration as to their final allotment. Soon thereafter, a Greek-Bulgarian anti-Turk military alliance was signed, in which no territorial issues were defined since both states desired the important Macedonian port of Thessaloniki, and Montenegro signed alliances with both Serbia and Bulgaria. Thus the Balkan League was constituted.

Meanwhile, the Ottoman state was harried by Italian attacks and in internal disarray. In May 1912 the Albanians rose against the Young Turks, and Ottoman military morale and strength began collapsing. By October 1912 conditions were ripe for the Balkan allies to commence war on the Turks.

Ignoring Russian pleas to wait, Montenegro declared war on 7 October, followed ten days later by the other Balkan allies. There was little doubt that the war was fought primarily to decide Macedonia's fate, but geography forced the Bulgarians, the easternmost of the allies, to focus their efforts in the wrong direction, against the main Turkish forces in Thrace, while their three allies faced mostly demoralized and weak enemy units in the west, in and around Macedonia. Serb forces easily overran Kosovo, occupied close to two-thirds of Macedonia, and then invaded Albania. Greek troops pushed into Epiros and southern Macedonia, occupying Thessaloniki over loud Bulgarian protests. The Bulgarians found themselves in a bloody contest for Edirne and received a gruesome foretaste of trench warfare in their assaults on successive Turkish fortified positions at Lüleburgaz and Çatalca. When Edirne fell to the Bulgarians in March 1913, only Istanbul itself and Shkodër, in Albania, remained of Turkey-in-Europe. In April an armistice was signed. In May the Treaty of London ended hostilities.

Dissension soon arose among the victorious Balkan allies over the disposition of conquered territories. The European Great Powers decided to create an autonomous Albania, which included areas originally ceded to Serbia in alliance treaties. In compensation, the Serbs demanded a larger share of Macedonia, to which the Bulgarians adamantly objected. The Bulgarians and Greeks were at loggerheads over possession of Thessaloniki. Smelling nationalist blood, the Romanians, who had remained neutral during the war, placed a bid for southern Dobrudzha, which had been in Bulgarian hands since 1878. Russia attempted to smooth the frictions among the allies but failed. In June 1913 Serbia and Greece concluded an anti-Bulgarian alliance to defend their zones of occupation in Macedonia against possible Bulgarian encroachment, and then won Montenegrin support. A Russian mediation initiative was ignored.

All three of the contentious allies transferred troops to the lines established in and around Macedonia. Border clashes between Bulgarians and their now belligerent former allies multiplied. Nationalist emotion in Bulgaria built to fever pitch. The army grew restless and demanded action or demobilization. The Bulgarian public, whipped up by the agitation of Macedonian immigrant groups who threatened to assassinate King Ferdinand and important members of his government if they did not act, clamored for war against both Greeks and Serbs. The military high command, which had hurriedly redeployed the bulk of the army from the eastern front facing the Turks to the western front facing Macedonia, assured Ferdinand that all was ready for decisive action. In late June the Bulgarians attacked Serb and Greek positions in Macedonia. It was a naive and foolish move.

Serbia and Greece immediately declared war on Bulgaria. Montenegro followed, and in July both Romania and the Ottomans did likewise. The Bulgarians were placed in an untenable military position and could offer only meager resistance to their enemies' concerted attacks. They were easily defeated by the Serbs and Greeks in Macedonia, while the Turks regained most of Thrace up to and including Edirne, and the Romanians captured southern Dobrudzha. In a little over a month the Second Balkan War was over. By maneuvering the Bulgarians into playing the role of aggressors, the anti-Bulgarian allies ensured that Bulgaria forfeited any sympathetic support among the international diplomatic community. In the Bucharest Treaty ending the war, Bulgaria was stripped of most gains won in the first war, including western Thrace and the port of Kavala, Edirne and most of eastern Thrace, and most new acquisitions in Macedonia, except for a slice in its northeast. Romania retained much of Dobrudzha, while Greece and Serbia divided the rest of Macedonia between themselves—the Greeks retaining Thessaloniki and the southern portions of the region; the Serbs acquiring the lion's share of the central and northern portions. The borders established for Bulgaria and Greece in 1913 have proved relatively stable up to the present day.

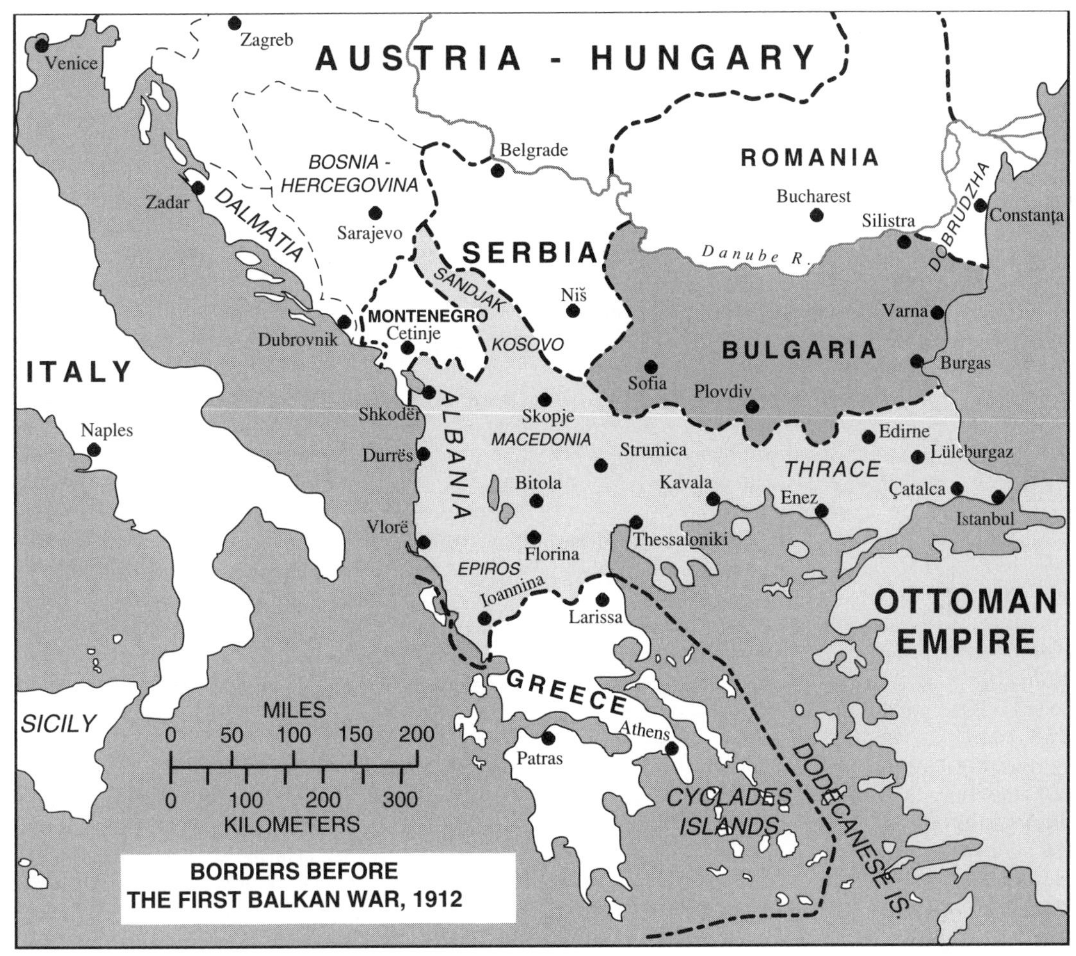

BORDERS BEFORE THE FIRST BALKAN WAR, 1912

BORDERS AFTER THE FIRST BALKAN WAR, 1912

BORDERS AFTER THE SECOND BALKAN WAR, 1913

Map 34: World War I in the Balkans

World War I was an accident. All of the Great Powers tried to head it off, yet war erupted. Austria-Hungary wanted to squelch the threat of Serb minority nationalism by thrashing Serbia decisively in a localized war. Once Habsburg troops invaded Serbia in July 1914, however, the weblike system of Central and Entente alliances precluded all efforts to stop the proliferation of military action. Russia's alliance with Serbia demanded military mobilization against Austria-Hungary. Germany then was obliged to mobilize in support of Austria-Hungary, which, in turn, forced France to follow suit according to its treaty with Russia. Unfortunately, military mobilization meant that combat could not be averted. Within a little over a month of the tragedy in Sarajevo, the European world found itself embroiled in total war.

World War I opened with an ineffectual bombardment of Belgrade by Austro-Hungarian artillery, followed by two unsuccessful Habsburg invasions of Serbia from Bosnia. Because of Belgrade's exposed position on the border with the enemy, Serbia's government moved to Niš, where, in the December Niš Declaration, it officially called for the unification of all Serbs, Croats, and Slovenes into an ill-defined "Yugoslav" state. A few Habsburg Serbo-Croat leaders, headed by the Dalmatian Croat Ante Trumbić, fled Austria-Hungary and founded the Yugoslav Committee in London for advancing the union of Habsburg South Slavs with Serbia and Montenegro.

The Ottoman Empire, at first neutral, entered the war in late 1914 on the side of the Central Powers. Bulgaria, smarting over its recent losses in the Balkan Wars, also remained neutral at the war's start, hoping to bargain its eventual participation to the side that would satisfy its nationalist territorial ambitions to the greatest degree. The Entente, ostensibly fighting the war in defense of Serbia, had little to offer Bulgaria, since the Serbs had no intention of relinquishing their hold on Macedonia, Bulgaria's primary nationalist claim, and both Greece and Romania, on whom Bulgaria also had territorial claims, were being wooed as potential Entente allies. Meeting Bulgaria's demands entailed forcing concessions from allies and friends alike. The Central Powers had no such concerns and freely granted the Bulgarians all that they desired since it cost them virtually nothing.

Bulgarian negotiations with the warring sides dragged on through most of 1915, during which Italy was enticed to forswear the Central Alliance (April 1915) and join the Entente with promises of Balkan Adriatic gains. A bungled anti-Ottoman Entente invasion of Gallipoli (April 1915-January 1916) convinced Bulgarian king Ferdinand to join the Central Powers for a combined Austro-Hungarian and Bulgarian assault on Serbia (October 1915). The outnumbered Serbs were crushed and forced to make a grueling winter retreat through the Albanian Alps to the Adriatic, where the survivors were evacuated by the Entente to Corfu. Serbia's ally Montenegro surrendered. Bulgaria occupied Macedonia and pushed toward Thessaloniki, where it was stopped by an Anglo-French force invited into Greece by Greek prime minister Elevtherios Venizelos, who, with Entente backing, displaced Greece's pro-German king Constantine (1913-17; 1920-22). An Entente front was established around Thessaloniki, to which the reorganized Serbian army was transferred, and by November 1916 the front line was pushed as far north as Bitola in southern Macedonia.

During 1915 the Entente succeeded in enticing Romania from the Central Alliance by promising it a blank check in winning its nationalist territorial claims in Transylvania. Immediately after declaring war against the Central Powers in August 1916, Romanian armies invaded Transylvania but were thrown back by German and Habsburg forces. Bulgarian troops then drove north through Dobrudzha into Romania. The Romanians were crushed by the united Central Powers forces and their army and government driven into Moldavia, where they tenuously retained a foothold around Iaşi thanks to timely Russian military intervention. Bucharest was occupied by the Central Powers and the bulk of Romania's oil- and grain-producing regions were captured. The failed attempt to win Transylvania cost the Romanians over half of their state.

During 1917 fighting was stalemated in the Balkans. The new Habsburg emperor Charles I (1916-22) questioned his continued participation in the war after defeats by Russia on the Eastern Front reduced Austria-Hungary to German satellite status. When the March Revolution took Russia out of the war and the Bolshevik star was on the rise, Charles, aware that defeat spelled disaster for his empire and fearful of social revolution, attempted to take Austria-Hungary out of the war. His efforts, though ultimately futile, spurred the Serbian government on Corfu and the London Yugoslav Committee to issue a joint Corfu Declaration (July 1917) calling for the independent unification of all South Slavs after the war to forestall a similar call by a Habsburg "Yugoslav Group" for unity within a postwar Habsburg state.

In the chaos in Russia following the Bolshevik Revolution (November 1917), Romania annexed Bessarabia (March 1918) and, left without Russia's military support, surrendered to the Central Powers (May 1918). As 1918 progressed, the Bulgarians, their morale shaken by German exploitation, grew less committed to the war. Exhausted Bulgaria was the first Balkan Central Powers ally to collapse. A concerted assault by the Entente forces from Thessaloniki (September) smashed the Bulgarian lines. Bulgaria's army disintegrated and French and Serbian forces raced deep into the Balkans' interior. In October both Bulgaria and the Ottomans called for armistices. Emboldened by events, Romania rejected its surrender (November) and invaded and captured Habsburg Transylvania against little resistance.

RUSSIA
AUSTRIA - HUNGARY
Vienna
Budapest
Ljubljana
SLOVENIA
Zagreb
Venice
Rijeka
CROATIA-SLAVONIA
BOSNIA-HERCEGOVINA
Sarajevo
DALMATIA
Split
Dubrovnik
Cluj
Iaşi
BESSARABIA
TRANSYLVANIA
BANAT
Brăila
ROMANIA
Belgrade
Bucharest
Constanţa
Danube R.
Silistra
SERBIA
Niš
Tŭrnovo
Varna
MONTE-NEGRO
Cetinje
Priština
BULGARIA
Sofia
Prizren
ITALY
Plovdiv
Skopje
Shkodër
ALBANIA
Naples
Strumica
Edirne
Durrës
MACEDONIA
Bitola
Kavala
THRACE
Istanbul
Vlorë
Florina
Thessaloniki
Gallipoli
Larissa
Ioannina
CORFU
GREECE
İzmir
SICILY
Athens
Patras
MILES
0 50 100 150 200
0 100 200 300
KILOMETERS
DODECANESE IS.
(To Italy)
CRETE
Territories involved in military operations during:
1914
1916
1915
1918
Thessaloniki Front, late 1916-late 1918

Map 35: The Post-Versailles/Lausanne Balkans, 1923

The victorious Entente Powers (specifically, Britain, France, and the United States) drew the 20th-century political map of the Balkans during negotiations at Versailles (Paris) ending World War I (1919-20). Because the United States had entered the war as an Entente ally with the stated goal of winning "national self-determination" for peoples lacking nation-states of their own (expressed by President Woodrow Wilson [1913-21] in his "Fourteen Points"), and American resources had proved decisive in the Entente's ultimate victory, the Entente allies' guiding principle at Versailles ostensibly was the creation of a new international order based on the "Fourteen Points." Britain and France, however, had entered the war to defend their national imperialist interests, and at the peace table they made cynical use of "national self-determination" in their mapmaking to punish the "losers" and reward the "winners." In drawing new state borders, claims of "winners" were acknowledged at the expense of equally valid, but disregarded, claims of nation-state "losers."

In the Balkans, where national emotions sparked by the 1878 Berlin Congress continued to run hot and nation-state border issues were paramount among all nationalists, the Versailles settlements exerted a lasting impact. The "winners" were the Serbs, Romanians, and Greeks, while the "losers" were the Bulgarians and the Ottomans.

During the last days of the war, Serbian forces from the Thessaloniki front overran all of prewar Serbia and constituted the only organized military presence in the Balkans' northwest. The Montenegrins and the former Habsburg Serbo-Croats and "Yugoslav Group" accepted Serbian leadership in a united South Slav state, which Serbian prince-regent Aleksandr I Karadjordjević (1921-34) proclaimed in Belgrade (December 1918). At Versailles, both the Corfu Declaration and the Belgrade Pronouncement were accepted as legal, and a new Kingdom of Serbs, Croats, and Slovenes (encompassing Serbia, Croatia, Slovenia, Slavonia, Dalmatia, Vojvodina, Kosovo, Bosnia, Hercegovina, and Macedonia) was recognized as a legitimate state. In nation-state and "national self-determination" (actually meaning British and French determination) terms, it was extremely artificial, as its official name demonstrated. Besides Serbs (including Montenegrins), Croats, and Slovenes, it supposedly represented Macedonians, Bosniaks (Muslims), and a smorgasbord of other ethnic groups scattered throughout its territories. Many were distinctly unhappy with "their" new state, especially many Slovenes, who had voted to remain with Austria in a 1920 plebiscite. In actuality, the kingdom primarily served the interests of the Serbs, who were its sole representatives at Versailles—their official "reward" from the Entente Powers. (See Map 36.)

Despite its rather miserable military performance during the war, and its somewhat duplicitous reentry at war's end to win easy spoils, Romania still emerged a "winner" at Versailles. Because Romania had surrendered to the Central Powers during the war, it was required to sign the Trianon Treaty with Hungary (1920) before its wartime annexation of Bessarabia officially was recognized. Getting Trianon signed proved a difficult process because Romania not only had occupied Transylvania at the war's end but had pushed on into Pannonian Hungary, destabilizing the post-Habsburg Hungarian Republic and ushering in a brief period of Bolshevik control in that state (1919). The Entente Powers pressured Romania to withdraw its troops to Transylvania so that a stable Hungarian government could be formed to sign the treaty. Trianon handed Transylvania, Banat, and part of Bukovina to Romania. Although the Hungarians signed, they never accepted Trianon's terms as final. (See Maps 37 and 38.)

Bulgaria received the punishment dealt all of the "losers" at Versailles. Following the elevation of King Boris III (1918-43) after Ferdinand's abdication at the end of the war, elections returned an Agrarian Union-led coalition government headed by Aleksandŭr Stamboliiski as premier (August 1919). Boris and Stamboliiski had no option but to sign the Neuilly Treaty (1919) and accept its punitive terms imposing a large war indemnity and severe military limitations. More nationalistically repugnant were the territorial terms, which stripped the state of all lands acquired since 1912. Southern Dobrudzha (won in 1916) was restored to Romania; the Serb-Croat-Slovene Kingdom received four small but strategic plots of land on Bulgaria's western border; and Bulgaria's portion of western Thrace that was won in the Balkan Wars (with access to the Aegean Sea) was given to Greece. Bulgarian requests for plebiscites to determine the involved territories' ethnic composition fell on deaf ears at Versailles. Resentment over the treaty's terms was widespread throughout Bulgaria, and the Bulgarians refused to consider it final. (See Map 39.)

Greece wanted far more than western Thrace from the Versailles settlements. Venizelos sought northern Epiros from Albania and all Greek-inhabited territories, including Istanbul, from the defeated Ottoman Empire. Secret wartime agreements parceling out Ottoman territories to various Entente allies—and Anglo-French, Italian, and, ultimately, Greek military interventions in Istanbul and Ottoman Anatolia—complicated the final treaty process. When the Entente tried to force the Ottomans to accept the Sèvres Treaty partitioning the empire, using Greek intervention forces operating out of Izmir as their muscle, a Turkish nationalist movement, headed by Mustafa Kemal (Atatürk), instigated both a civil war against the Ottoman government in Entente-occupied Istanbul and a war of national liberation against the Greek invaders. By 1923 Kemal deposed the last sultan, Mehmed VI (1918-22), ending the Ottoman Empire; forced the Greeks out of Anatolia; and rendered Sèvres a dead letter. Kemal's successes convinced the Entente and Greece (reluctantly) to sign the Lausanne Treaty (1923), which preserved Anatolia and a portion of eastern Thrace (including Istanbul and Edirne) for a newly constituted Turkish nation-state. (See Map 40.)

CZECHOSLOVAKIA
POLAND
SOVIET UNION
Danube R.
Vienna
Bratislava
AUSTRIA
Budapest
Klagenfurt
HUNGARY
MARAMUREŞ
BUKOVINA
BESSARABIA
SLOVENIA
Trieste
Venice
Ljubljana
Zagreb
Rijeka
CROATIA
CRIŞANA
Cluj
TRANSYLVANIA
MOLDAVIA
Iaşi
Kishinev
Alba Iulia
ROMANIA
SLAVONIA
VOJVODINA
Timişoara
BANAT
Novi Sad
K. OF SERBS, CROATS AND SLOVENES
BOSNIA
HERCEGOVINA
DALMATIA
Belgrade
WALLACHIA
Ploieşti
Bucharest
Craiova
Silistra
DOBRUDZHA
Zadar
(To Italy)
Sarajevo
SERBIA
Danube R.
ITALY
MONTENEGRO
Niš
Varna
Dubrovnik
Cetinje
Sofia
Rome
KOSOVO
BULGARIA
Skopje
Shkodër
Plovdiv
Naples
MACEDONIA
Edirne
Durrës
Tiranë
Ohrid
THRACE
Kavala
Istanbul
ALBANIA
Thessaloniki
Vlorë
EPIROS
Ioannina
CORFU
GREECE
TURKEY
MILES
0 50 100 150 200
0 100 200 300
KILOMETERS
SICILY
İzmir
Athens
DODECANESE IS.
To Italy
CRETE
Borders of "New" states at Versailles, 1919-1920
Region of Klagenfurt plebiscite zone, 1920

Map 36: Yugoslavia,, 1929-1941

The premise on which the Belgrade Pronouncement was issued, and the basis for the subsequent international recognition of a unified South Slav state at Versailles, was the wartime joint Corfu Declaration (1917) of the Serbian government-in-exile and the London Yugoslav Committee. (See Map 34.) The agreement between the Serbian prime minister and nationalist Nikola Pašić and the London Committee's Ante Trumbić, a Serb-leaning Dalmatian Croat, outlined the organization of an independent postwar Yugoslav state. Pašić agreed to Croat demands for a constitutional monarchy responsible to a democratically elected national assembly as the governing framework for a nationally confederated state. He did so, however, only because of political circumstances: The Serbs lacked the customary Russian support for their "Greater Serbia" claims (the revolution had overthrown the *tsar*); the Americans, newly involved in the war, favored a Croat-inspired federalist Yugoslav idea; and Serbia was occupied by enemy Central Alliance forces.

Despite Pašić's apparent compromise in the agreement, the Serbs continued their ultranationalist "Greater Serbia" Yugoslav approach for the duration of the war, much to the apprehension of the Croat and Slovene Yugoslav federalists. As the fighting ended in 1918 and the Habsburg Empire disintegrated, the Serbian military presence in the territories encompassed by the desired postwar Yugoslav state forced the various "Yugoslav" nationalists of Serbia, Croatia, Montenegro, and Slovenia to accept Serbian leadership, culminating in international recognition of a new South Slav state at Versailles. (See Map 35.)

Doubts regarding the sincerity of Serbia's promises to accommodate its new national partners surfaced quickly. The Croats and those Slovenes who decided to stay in the new state soon found that concerns about ties to the Serbs were valid. The Serbs moved to organize the state as a centralized Serbian nation-state under strong royal control by declaring Serbian prince Aleksandr Karadjordjević king-regent. Both the Croats and Slovenes envisioned the state's political structure in more liberal-democratic and federalist terms that would grant them significant national autonomy. Neither Croats nor Slovenes were happy to think that the Serbs—who in their minds were cultural inferiors, whose royal house stemmed from glorified, illiterate pig farmers a century earlier—intended to play the role of dominant partner in a highly centralized state. That Serbs dominated the new state in every way, controlling the top government ministries and offices, the military's officer corps, and the police with the support of Aleksandr's army, boded ill for the future of Croats, Slovenes, and other non-Serbs within the kingdom.

In 1921, the Serbs forced through a strongly centralized state constitution that essentially copied Serbia's prewar monarchical document, solidifying a national breach between Serbs and Croats. In 1924, after ill-timed Croat boycotts of elections and parliamentary proceedings initially failed to win political concessions from the Serbs, the Croats, led by Stjepan Radić, entered the political process and won seats in both the national assembly and royal cabinet. After Radić was shot fatally in the assembly's chamber by a radical Serb in 1928, the Croats withdrew their political participation in Belgrade, demanded a new federal constitution, and established their own separatist government in Zagreb, Croatia's capital. In response, the next year (1929) Aleksandr proclaimed a royal dictatorship, dissolved the nationalist Croat Peasant party, and arrested Croat leaders. The state was renamed Yugoslavia and administratively reorganized into geographically defined *banovinas* in an effort to eliminate all traces of historical national associations. A new constitution was issued in 1931 that essentially rigged any political participation in favor of the Serb-dominated royal government, and Croat and Slovene oppositional leaders continued to be arrested.

Out of that situation arose among the Croats an extremely radical, ultranationalist terrorist organization—the *Ustaše*, which forged close relations with outside Hungarian and Italian fascist leaders and internal Bulgaro-Macedonian terrorists. In 1934, from its headquarters in Hungary, the *Ustaše* masterminded the assassination of King Aleksandr in Marseilles. Less radical Croats offered to cooperate with Prince Pavel, the regent of the dead king's young successor Petr II Karadjordjević (1934-41), but he reneged on his promised political concessions to them, once again tilting Croat opposition back toward the extremists. This trend was reinforced in 1937, when Pavel's regency government dropped a project that, in accordance with a concordat made with the Vatican, would have granted wider privileges to Roman Catholics—Croats and Slovenes—because of widespread disturbances instigated by Orthodox groups and Serb radicals who opposed any such arrangement.

Rising Croat opposition, coupled with a growing democratic movement among more farsighted Serbs, led to serious discussions with the royal government in 1938 and 1939 regarding conciliatory moves to relieve the pressures building in the state. Royal dictatorship was ended. Fearful of rising Nazi Germany, Regent Pavel drew closer to Nazi strongman Adolf Hitler (1933-45) and tried to settle Yugoslavia's internal divisions. Pavel granted the Croats an autonomous territory (the *Sporazum)* within the state (1939) and offered their leader, Vladko Maček, a vice-premiership. Extremists on both sides remained unhappy, and the Slovenes and Bosnian Muslims demanded similar autonomy. In March 1941 progress came to an end. Serbian nationalists, who equated Germans with their former Habsburg nemesis, staged a successful military coup against Pavel and his pro-German policies and placed the youthful Petr on the throne. Ten days later, Hitler invaded Yugoslavia. (See Maps 42 and 43.)

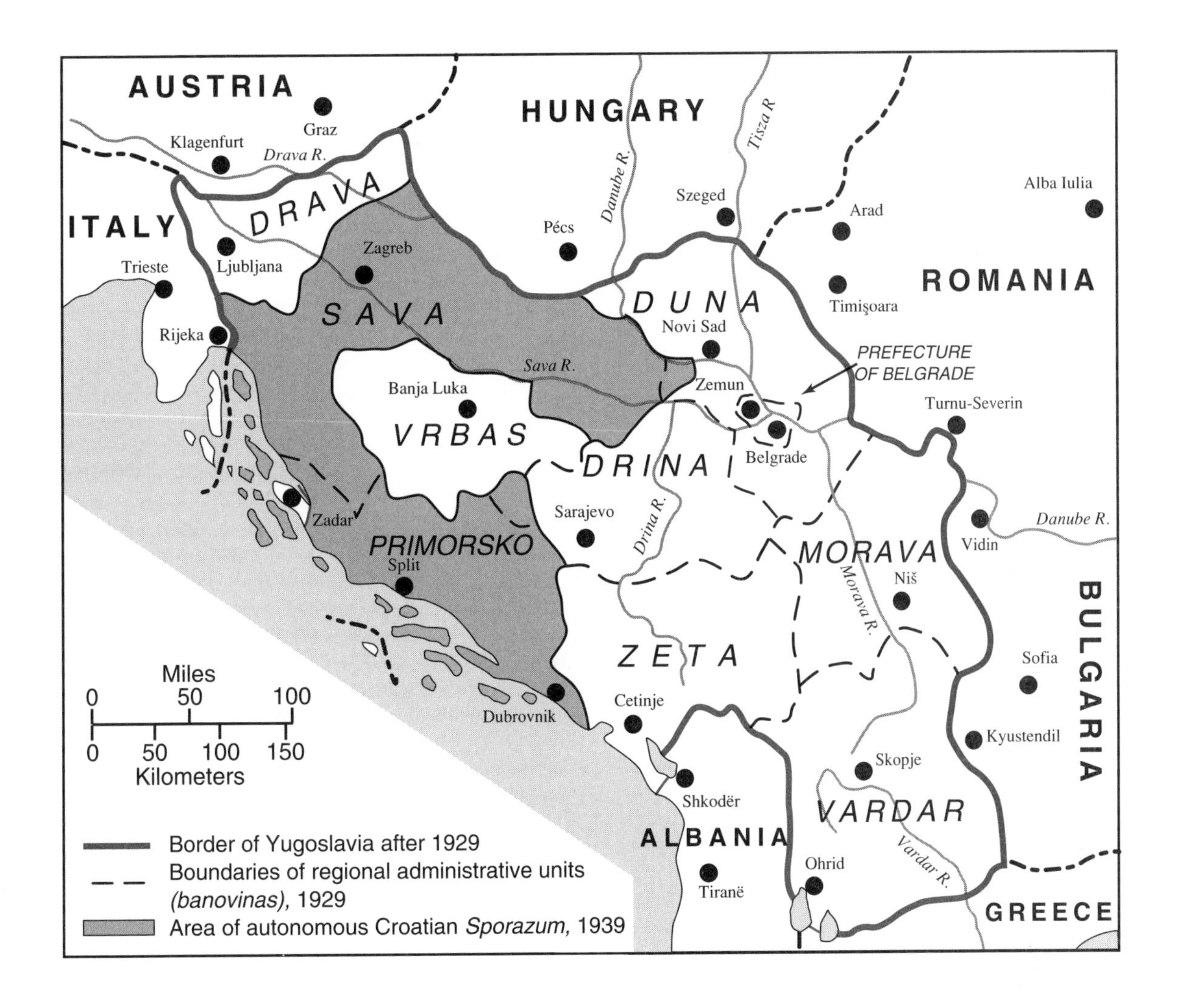
AUSTRIA
HUNGARY
ITALY
ROMANIA
BULGARIA
ALBANIA
GREECE
Graz
Klagenfurt
Drava R.
Danube R.
Tisza R
Szeged
Arad
Alba Iulia
Pécs
DRAVA
Zagreb
Ljubljana
Trieste
SAVA
DUNA
Timişoara
Novi Sad
Rijeka
PREFECTURE OF BELGRADE
Sava R.
Zemun
Banja Luka
Turnu-Severin
VRBAS
DRINA
Belgrade
Zadar
Sarajevo
Drina R.
Danube R.
Vidin
PRIMORSKO
MORAVA
Split
Niš
Morava R.
ZETA
Sofia
Miles
0 50 100
0 50 100 150
Kilometers
Dubrovnik
Cetinje
Kyustendil
Skopje
Shkodër
VARDAR
Vardar R.
Ohrid
Tiranë
Border of Yugoslavia after 1929
Boundaries of regional administrative units (banovinas), 1929
Area of autonomous Croatian Sporazum, 1939

Map 37: Post-Trianon Romania, 1920-1938

Throughout the 18th century, the thrones of the Wallachian and Moldavian principalities were sold by the Ottomans to members of wealthy Greek Phanariote families, who ruled their domains in Byzantine autocratic fashion, surrounded by dependent landowners, and subject only to the whims of the Ottoman sultans. Under Phanariote rule, one of the most oppressive and inequitable social systems in Eastern Europe emerged, with a small privileged elite of powerful magnates lording over the majority of poor, enserfed peasants. During the 19th century, Russian presence north of the Danube led to the establishment of a Russian protectorate, essentially replacing Greek with Russian control in 1829.

Romanian nationalism originated among Transylvanian Romanians in the Habsburg Empire and filtered over the Carpathians into the principalities by means of economic and cultural (Orthodox) links. It exploded in the revolutions of 1848. The Principality Romanians, under heavy French influence, exhibited a highly aristocratic and essentially feudal form of nationalism that was concerned with blotting out the Greek Phanariote legacy of their Ottoman past and looked to build a "Greater Romania," while the Transylvanian Romanians, whose nationalism was based on a policy of social reform for an egalitarian society, struggled to gain recognized parity with other nationalities and social reform within the Habsburg Empire.

The Romanian revolutions in 1848 resulted in successive periods of Russian and Austrian occupation of the principalities until 1857. At a European Great Powers conference in Paris (1858) following the Crimean War, Wallachia and Moldavia were permitted to establish common institutions under separate princes. Alexander Cuza (1859-66) managed to win election as prince in both principalities, and by 1862 the union was recognized as a single country, Romania. Cuza squelched all internal opposition by strengthening the legal powers of the princely office, but his efforts to institute social reforms earned him the wrath of the powerful magnates, who forced him to abdicate in 1866. His replacement was a German, Prince Carol I, who won full independence from Russia in 1878 as a reward for Romanian participation in the 1877-78 Russo-Turkish War, though at the expense of ceding Bessarabia to Russia. (See Map 27.) In 1881 Carol proclaimed himself king. His reign was characterized by internal disturbances springing from the inequitable social division between the wealthy landowners and the poor peasants, which resulted in long periods of martial law.

Carol I started agitation among the Transylvanian Romanian nationalists for union with his kingdom before the outbreak of World War I. By joining the Central Powers in 1883, he hoped to gain Transylvania in return. Thereafter, he supported the Romanian "Greater Romania" nationalists' public anti-Hungarian attacks aimed at gradually separating Transylvania from Hungary through the efforts of its native Romanian nationalists. Although the Hungarian government did its best to suppress Romanian nationalist agitation, Habsburg crown prince Francis Ferdinand, who despised and distrusted the Hungarians, secretly promised Carol Transylvania for Romania if Carol agreed to join his country to the Habsburg Empire. By the time war erupted in 1914, "Greater Romania" agitation among the Transylvanian Romanians was highly developed, but largely unsuccessful.

The collapse of Austria-Hungary in World War I led to the union of Transylvania and Romania, including large Hungarian and German minority populations. (See Map 35.) In late 1918, taking advantage of the Russian Bolshevik Revolution, Romania also annexed Bessarabia, which had a mixed population of Romanians, Turks, Bulgarians, Ukrainians, and Russians. The Treaty of Trianon validated both of those territorial acquisitions. The Romanians then implemented programs that would weaken the minorities in their newly enlarged state and strengthen themselves. Non-Romanian governing institutions were eliminated and minority officials systematically were weeded out of their posts. Public schools became tools for Romanianizing the minorities, while minority church and private schools were either seized by the government or closed on the most specious of pretexts. Beatings and imprisonments of non-Romanians became commonplace.

Much of the rabid antiminority policy took place behind a screen of liberal legalism. The 1923 constitution was a model of liberal-democratic ideals. The law was good; its enforcement was not. Romanian officials bullied the minorities living in their districts and showed utter contempt for the laws that they were sworn to uphold. The Agrarian Reform of 1920 also was an outwardly progressive policy that was used to discriminate against non-Romanians. While certainly needed in the former principalities, where traditional feudal relationships still existed, it also served to deprive minorities of their land, which supported the religous and educational activities that bolstered minority ethnonational awareness. Additionally, the the minority problem was aggravated by rampant anti-Semitism among the peasantry, who viewed the influx of refugee Jewish land stewards from Poland and its Catholic-charged nationalist euphoria as another form of landowner oppression.

When royal power declined under King Ferdinand (1914-27), the facade of liberal democracy proved unable to exert stable government because of social unrest over land reform, anti-Semitism, and minorities. Political assassinations became the norm in Romanian politics. King Carol II (1930-40) was forced to deal harshly with the "Iron Guard" (officially the "Legion of the Archangel Michael"), one of Europe's earliest authentic fascist movements, by establishing a royal dictatorship and ordering the murder of the Guard's leaders, including founder Corneliu Codreanu, in 1938.

CZECHOSLOVAKIA
POLAND
USSR
HUNGARY
YUGOSLAVIA
BULGARIA
BLACK SEA
Dniester R.
Southern Bug R.
Tisza R.
Someş R.
Prut R.
Mureş R.
Olt R.
Danube R.
Argeş R.
Sava R.
MARAMUREŞ
BUKOVINA
BESSARABIA
CRIŞANA
TRANSYLVANIA
MOLDAVIA
BANAT
WALLACHIA
DOBRUDZHA
Chernivtsi
Hotin
Satu-Mare
Debrecen
Oradea
Iaşi
Kishinev
Odessa
Szeged
Cluj
Tîrgu-Mureş
Arad
Alba Iulia
Timişoara
Sibiu
Braşov
Galaţi
Izmaïl
Brăila
Ploieşti
Belgrade
Turnu-Severin
Bucharest
Craiova
Constanţa
Giurgiu
Silistra
Vidin
Ruse
Varna
Pre-1918 Romania
Romania's Trianon acquisitions, 1920
MILES
0 50 100
0 50 100 150
KILOMETERS

Map 38: The Transylvanian Question

While Hungarian nationalists ardently sought to reclaim all territories and populations torn from "historic" Hungary by Trianon, no loss was more galling or more fervently disputed than that of Transylvania. Hungarian attempts to gain international redress on this issue began at Versailles in the negotiations leading up to the treaty and were actively continued before the Minorities Question Section of the League of Nations' Secretariat and through the national and international print media. The Hungarians' unwillingness to let their Transylvanian cause subside from international notice forced Romanian nationalists to respond in kind. Although from 1919 until 1940 both the Hungarians and Romanians poured mountains of statistical data relating to demographic, economic, and political issues into supporting their respective arguments, the heart of both sides' cases justifying their conflicting claims on Transylvania was historical. The incessant public dispute between them became known as the "Transylvanian Question." It lasted as an open diplomatic sore until 1940, when Hitler attempted to force a compromise solution on the two sides; thereafter, the end of World War II and the subsequent submergence of both contending parties beneath the tide of communism hid the conflict below the surface of Soviet-imposed international socialist brotherhood.

Transylvania had been incorporated into the medieval Hungarian state in the 11th century. During the 16th and 17th centuries it had enjoyed a golden age of quasi-independence as an Ottoman client and Hungarian bastion of anti-Habsburg Protestantism, and in the Hungarian Revolution of 1848-49 it had served as the final fortress of the revolutionaries against the forces of both the Habsburgs and the Russians. To Hungarian nationalist minds, it was unthinkable that a region of such historical national importance should be in the hands of Romanians. Through continuous representations before the League of Nations and its minority rights commission in Geneva, voluminous publications in Western languages, and constant political agitation both within and outside of Hungary between 1920 and 1939, the nationalist Hungarians kept the Transylvanian issue burning on the stage of international public opinion. The Hungarians agitated so rabidly for a revision of Trianon that their neighbors, all of whom had received slices of Hungarian territory in the treaty, formed the Little Entente to protect themselves from possible Hungarian efforts to revise Trianon by force. That only inflamed Hungarian nationalism further.

Romania was forced to counter the Hungarian historical arguments for control of Transylvania. The Hungarians claimed that Transylvania was uninhabited when their ancestors conquered the region in the 11th century. For that reason Hungarians from Pannonia were settled there and outside colonists—Székelys (ethnic relatives of the Hungarians) and Saxon Germans—were invited in to help protect and economically develop the region. According to the Hungarians, Romanians only entered Transylvania in significant numbers (from the Balkans to the south) starting in the late 12th and early 13th centuries. As a lowly peasant people, the Romanians remained of little political or cultural account compared to the original Hungarian, Székely, and Saxon populations, who shouldered the burdens of administering, protecting, and developing the region.

The Romanians claimed the region by right of demographic majority—55 percent of the region's population at the time of Trianon was Romanian—and by right of historical possession. According to the Romanians, their ancestors were the ancient Daks, who controlled Transylvania, were conquered by the Roman emperor Trajan (98-117), and placed under Roman authority for 160 years, during which time they were Latinized. When Rome withdrew from the region in 271, the Romanized Daks remained behind, finding refuge in the mountains from the tidal waves of Germanic and Turkic migrations that swept through the region from the 3rd through 9th centuries. They reemerged in the Transylvanian lowlands by the 11th century, when they were conquered by Hungarian intruders. The Romanians' theory of their origins thus reduced all the other nationalities in Transylvania to historical interlopers in a native Romanian homeland.

In 1939 both Hungary and Romania found themselves allied with Germany—the Hungarians because of a sense of nationalist-revisionist common cause with the Germans and the Romanians out of economic necessity. (See Map 42.) Hitler needed them both—the Hungarians to ensure his political dominance in Eastern Europe and to protect his southern flank and the Romanians to provide necessary oil and manpower resources—for his planned future military operations against the Soviet Union. In 1940, when the Hungarian-Romanian conflict over Transylvania threatened to explode into warfare, Hitler attempted to impose a solution to the problem. In the Second Vienna Award (30 August 1940) Hitler gave the northern 40 percent of Transylvania, including the Székely region in the extreme southeast, with a population of some 2.5 million people (52 percent of whom were Hungarians), to Hungary.

Although Hitler considered the matter settled, and the two antagonists had no choice but to accept his dictate, the Vienna solution solved nothing. Neither the Hungarians nor the Romanians considered the award anything other than a temporary settlement that would be worked out once Germany won World War II. Its immediate result was the disaffection of both allies. Unfortunately for the Hungarians, the award was nullified by German defeat, and Romania defected at the last moment to the anti-German allies at the end of 1944 and received all of Transylvania—and more—as a reward. (See Map 44.)

HUNGARY
BUKOVINA
Satu-Mare
MARAMURES
Debrecen
Tisza R.
Someş R.
Bistriţa
MOLDAVIA
Oradea
Criş R.
CRIŞANA
Cluj
Tîrgu Mureş
Szeged
Arad
Alba Iulia
Blaj
Mureş R.
Timişoara
Tisza R.
Hunedoara
Sibiu
Braşov
BANAT
Olt R.
YUGOSLAVIA
WALLACHIA
Danube R.
Belgrade
Turnu-Severin

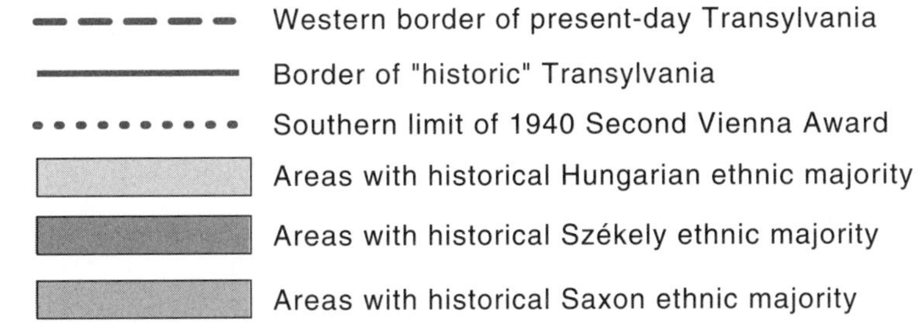
Western border of present-day Transylvania
Border of "historic" Transylvania
Southern limit of 1940 Second Vienna Award
Areas with historical Hungarian ethnic majority
Areas with historical Székely ethnic majority
Areas with historical Saxon ethnic majority

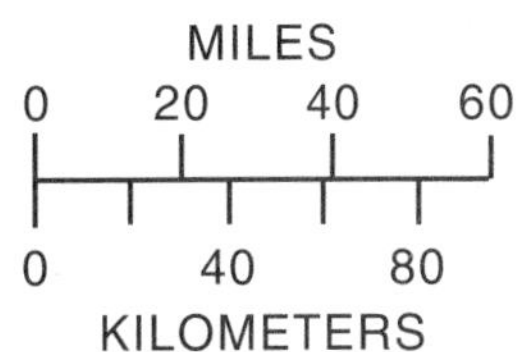
MILES
0 20 40 60
0 40 80
KILOMETERS

Map 39: Bulgaria, 1919-1940

Bulgaria was the chief Balkan state "loser" at Versailles. Besides the hobbling territorial losses, reparations payments, and military restrictions imposed by the Neuilly Treaty, the state was burdened with some 250,000 refugees from Macedonia and Thrace, who caused overcrowding and concomitant political and social pressures for the postwar Agrarian Union government headed by Aleksandŭr Stamboliiski. As leader of Europe's first peasant-oriented government, Stamboliiski represented a symbolic alternative to Soviet leader Vladimir Lenin's (1917-23) Bolshevism for many postwar-era European statesmen, who hoped that his "Orange" movement might stymie the "Reds'" spread.

Stamboliiski espoused a radical policy of peasant empowerment aimed at building an egalitarian society enjoying modern material benefits such as clean habitations, paved streets, good water, proper sanitation, universal primary and secondary education, and cheap, abundant necessities. All state lands and larger private estates (there were no great ones because there were no aristocratic landowners) were redistributed; a heavy progressive income tax was levied on all but the peasants; compulsory labor service replaced military service for youths; the middle class's governmental role was reduced; rural education was expanded and curricula modified to include increased vocational and technical training. Stamboliiski's Agrarians alienated the commercial and professional classes, who often were targeted as threats to their egalitarian goals. The Bulgarian Communists considered the Agrarian Union their sociopolitical enemy, while traditional political parties chafed over being discredited for past political ineffectiveness.

Unfortunately for the Agrarians, Stamboliiski's governing approach was authoritarian and ruthless. He cleaned house in the government, replacing most previous ministers with inexperienced peasant leaders who often were vulnerable to corruption. Elections were manipulated and political opposition muzzled through rigid press censorship and police crackdowns.

Stamboliiski's foreign policy ultimately proved literally fatal for him. He personally was uninterested in national territorial expansion, which so pleased the Entente victors that Bulgaria was the first defeated state admitted to the League of Nations (1920) and reparations were reduced (1923). He tried to dampen the Macedonian issue by developing friendly relations with the Yugoslav Kingdom and clamping down on IMRO, stirring up the intense hatred of IMRO leaders and nationalistic army officers. They staged a coup (June 1923), in which Stamboliiski brutally was murdered. Effective central authority evaporated for a time under a reactionary government. Nationalist agitation increased to dangerously anarchistic levels, and IMRO stepped up terrorist raids into Yugoslav and Greek Macedonia.

Abetted by the general frustration over the Neuilly Treaty's terms, reports of Yugoslavia's and Greece's bad treatment of Macedonian Slavs in their Macedonian holdings, and the large Macedonian immigrant population's bitterness over the postwar loss of Macedonia, IMRO played a leading role in Bulgarian politics for a decade following the coup. It was entrenched in southwestern Bulgaria, which it essentially governed for all practical purposes as a small Macedonian state within Bulgaria. IMRO's increasing terrorist incursions led Yugoslavia and Greece to fortify their borders with Bulgaria. IMRO, however, was wracked by internal dissension among "annexationists," "autonomists," and newly emerging IMRO Communists, which degenerated into internal civil war. Between 1924 and 1934 some 400 persons died. Calmer heads within Bulgaria eventually tired of the violence. Realizing that normalizing relations with Yugoslavia and Greece was necessary for future Balkan peace, a group of reserve officers and progressive intellectuals staged a coup (1934) that established a one-year dictatorship. As one of their first acts, IMRO was suppressed.

By the time of its demise, IMRO no longer was a true revolutionary organization but a racketeering-gangster operation extorting money from the Macedonian immigrants in southwestern Bulgaria. It was heavily involved in illegal drug manufacturing and smuggling and operated numerous opium refineries. In the end, IMRO espoused no concrete program other than the retention of its wealth and power through empty slogans and naked violence.

By 1935, a decade of IMRO gangsterism, ballooning problems caused by Neuilly's terms, an agrarian economy that increased the state's debt and dependency on the world market, a flood of Macedonian and Thracian refugees, continuing border incidents with Greece and Yugoslavia, and rising Communist activity (culminating in a failed bombing attack on King Boris in 1925, after which the Communist Party was declared illegal), had reduced Bulgaria's political situation to anarchy. Boris attempted to stem the chaos and reassert central royal control by disbanding the military dictatorship (1935) and inaugurating a royal dictatorship (1936), which ultimately proved unpopular.

The 1934 Balkan Entente among Yugoslavia, Greece, Romania, and Turkey was an effort by Bulgaria's neighbors (who had benefited territorially at Bulgaria's expense in the Neuilly Treaty) to stymie Bulgarian revisionism. Boris, however, managed to conclude formal agreements with Yugoslavia (1937) and Greece (1938) that somewhat lessened Balkan tensions. Bulgaria was permitted to rearm, but that process already had started with German support. German influence on the state was growing, although Bulgaria was not wholly committed to Nazi Germany. A Bulgarian National Socialist organization set up by Germany was disbanded (1938), and Boris readily accepted a large Anglo-French loan to support his rearmament program. As world war loomed in 1939, efforts were made to bring Bulgaria into the pro-Anglo-French Balkan Entente, but they were frustrated by Boris's demand to negotiate a revision of Neuilly's territorial terms. Thereafter, Bulgaria moved closer to Germany.

ROMANIA
Bucharest
Craiova
Giurgiu
Silistra
Constanţa
DOBRUDZHA
YUGOSLAVIA
Vidin
Danube R.
Ruse
Nikopol
Svishtov
Montana
Vratsa
Pirot
Shumen
Lovech
Varna
Yantra R.
Iskŭr R.
Tŭrnovo
Tsaribrod
Sofia
Pernik
Zlatitsa
Karlovo
Tundzha R.
Sliven
Bosilegrad
Kyustendil
Samokov
Stara Zagora
Burgas
BLACK SEA
Kriva Palanka
Gorna Dzhumaya
Plovdiv
Maritsa R.
Struma R.
Edirne
Smolyan
Strumica
Nevrokop
Maritsa R.
Lüleburgaz
MACEDONIA
Mesta R.
TURKEY
Serres
Kavala
GREECE
THRACE
Thessaloniki
AEGEAN SEA
SEA OF MARMARA
Bulgaria, 1919-1940
Territories lost by Neuilly Treaty, 1919
Miles
0 20 40 60
0 20 40 60 80
Kilometers

Map 40: Greece, 1923-1941

In May 1919, Greek prime minister Venizelos gained permission from the Entente Powers at Versailles for Greece to land in Ottoman western Anatolia and occupy Ízmir. The atrocities perpetrated on the local Turkish population by the Greeks sparked the outbreak of Mustafa Kemal's militant Turkish nationalist movement and eventually checkmated the Sèvres Treaty. Outright war erupted between the Turkish nationalists and Greece, which dragged on for some four years, ending with the Greek military expedition's utter defeat and the collapse of Greek nationalist dreams for acquiring the Anatolian coastline (with Izmir) and most of Thrace. At Lausanne (1923), Greece was forced to recognize the permanence of the new Turkish state and accept an exchange of respective minority populations. (See Map 35.)

The postwar Anatolian adventure and the subsequent Lausanne Treaty had lasting repercussions for Greece. After the death of King Alexander (1917-20) in the midst of the intervention (1920), deposed pro-German king Constantine returned, causing Greece to forfeit Entente support. Constantine, in turn, was ousted because of his military failure in Anatolia, and King George II (1922-23; 1935-47) was installed in his place. George was unable to deal with the problems stemming from the Lausanne population exchange and was overthrown for a republican government headed by Venizelos (1923).

The compulsory exchange of minorities called for at Lausanne involved some 1.3 million Greeks expelled from Turkey and 480,000 Muslims removed from Greece. The exchange was a mixed blessing for Greece. Its Macedonian territories were strengthened ethnically—numerous non-Greeks left, and most Greeks repatriated from Turkey were settled there to reinforce the Greek presence. Remaining non-Greeks were subjected to forced Hellenization. Attempts were made to eradicate permanently any non-Greek ethnocultural presence in Greek Macedonia, singling out Slav inhabitants—officially termed "Slavophone Greeks"—for particular discriminatory attention. Unfortunately, the Greek government and nationalists never felt completely certain that their Hellenization policy in Macedonia succeeded, and a sense of paranoia regarding a "Slav threat" to Greece's national hold on its northern territories persisted. Greek insecurity over Macedonia was intensified by continuous incursions into the region by Bulgarian-based IMRO terrorists during the 1920s and 1930s.

The population exchange also raised serious socioeconomic issues. The influx of more than a million immigrants into an overcrowded state was costly and socially explosive. There was not enough land to satisfy the immigrants' overall needs, leading to reduced average landholdings, increased agrarian impoverishment, and intensified general discontent. Both native mainland Greeks and Anatolian immigrants experienced a sense of culture shock, and mutual prejudices emerged.

National frustration over the postwar Anatolian adventure and the socioeconomic impact of the exchange were reflected in Greece's political sphere. A sense of national defeat and abandonment permeated politics, resulting in a succession of governments exhibiting various institutional forms—the military junta and puppet monarchy of George II (1923-24), a constitutional republic (1924-35), and the constitutional monarchy of a restored George II (1935-36)—ultimately culminating in General John Metaxas's military dictatorship (1936-41) with the blessing of King George. Western-style liberal-democratic institutions proved unable to cope with the interwar period's national, social, and economic pressures, and the Greeks resorted to the authoritarianism rooted in their Orthodox cultural traditions.

Even during the republic's 11-year existence, one individual—Venizelos—dominated affairs. As the premier for a series of presidential nonentities (1928-33), he manipulated electoral laws to ensure republican parliamentary majorities against royalist opposition. To his credit, Venizelos realistically used his power to address Greece's domestic and foreign problems, conducting progressive agrarian policies, fostering industrial development, and expanding Greece's merchant marine. Venizelos ended Greece's international isolation, called the first of three Balkan Conferences to improve the Balkan states' mutual political and economic relations, and began mending Greek-Turkish relations. He no longer held office when the Balkan Entente was signed by Greece, Yugoslavia, Romania, and Turkey in 1934, but his diplomatic efforts laid much of the preparatory groundwork for that alliance.

In 1935, the royalists restored King George II to the throne. George appointed retired general John Metaxas premier (1936), who convinced the king that "strong government" was necessary. Metaxas then staged a coup, with George's acquiescence and the army's support, and eventually assumed the premiership for life (1938), ending all liberal-democratic pretense. His dictatorship was both nationalist and populist—he called his governing ideology the "Third Hellenic Civilization"—emphasizing conservative Orthodox values and the vernacular Greek language.

Metaxas made efforts to win popular support by nationalizing the industrial sector and initiating social benefits. Numerous agricultural debts were canceled and low-interest agrarian loans made available. The merchant marine again was expanded, and a huge public works program emphasizing military rearmament was implemented. In foreign affairs, Metaxas maintained close relations with Turkey and adhered to the Balkan Entente. He also, however, laid Greece open to German economic penetration in the late 1930s. Germany's economic presence did not eclipse Greece's reliance on British and French support. As a new world war threatened in 1938, Metaxas openly expressed his Anglo-French sympathies, and when Italian Fascist leader Benito Mussolini (1922-43) occupied Albania in 1939, Metaxas accepted British and French guarantees of Greece's territorial integrity in return for resisting Italian aggression.

Shkodër
Skopje
Maritsa R.
Plovdiv
Struma R.
Mesta R.
YUGOSLAVIA
BULGARIA
BLACK SEA
Durrës
Tiranë
Ohrid
MACEDONIA
Vardar R.
Edirne
Serres
Kavala
THRACE
Istanbul
ALBANIA
Florina
Kastoria
Thessaloniki
Alexandroupolis
SEA OF MARMARA
Vlorë
Aliakmon R.
Mt. Athos
Bursa
Ioannina
Trikala
Larissa
AEGEAN SEA
TURKEY
Arta
Lamia
Arakhthos R.
Khalkis
İzmir
Thebes
Athens
Patras
Corinth
Pirgos
Navplion
Tripolis
Sparta
IONIAN SEA
CYCLADES IS.
DODECANESE IS.
(To Italy)
RHODES
Miles
0
50
100
0
50
100
150
Kilometers
Iraklion
CRETE
Greece, 1923-1941
Territories gained in 1919 but lost by Lausanne Treaty of 1923
MEDITERRANEAN SEA

Map 41: Albania, 1921-1939

An independent Albanian nation-state was created at the London Conference (1913) ending the First Balkan War, placed under joint Great Power protection, and given German prince William of Wied (1914) as ruler. The new state suffered from unfixed borders, which failed to include all territories claimed by Albanian nationalists (particularly Kosovo and northwestern Macedonia). Combined with the Albanians' endemic tribal, social, and linguistic disunity and general distrust of outsiders, this made William's reign brief. He fled the state early in World War I, and Albania was left prey to military occupations by both sides throughout the war. In 1917 Italy claimed Albania as a protectorate and only relinquished that assertion (1920) to focus its efforts on gaining Dalmatia and Istria at Versailles. At war's end, an Italian-supported national assembly sat in Durrës (1919), and anti-Italian Albanian nationalists attacked the occupying Italians and created a government in Tiranë (1920), headed by local clan chieftain Ahmed Zogolli. A Tiranë-inspired anti-Italian uprising pushed out the Italians (September 1920), solidified Zogolli's leadership of the nationalists, and earned international recognition for an independent Albania generally within its 1913 borders (1920). Albania became the only European state with a Muslim majority population (approximately 70 percent, with 20 percent Orthodox and 10 percent Catholic), and its governing institutions leaned heavily on Ottoman traditions.

The Tiranë government theoretically was a regency, but Zogolli, a traditional authoritarian centralist, dominated its affairs by controlling important ministries and adeptly maneuvering among various political factions. True to form, the new government was rent by factionalism, which coalesced roughly along two lines. The rich central and northern landowners (Ghegs) wanted to retain existing Ottoman legislation and opposed land reform, while southern middle-class reformers (Tosks), led by Zogolli and Bishop Fan Noli (a former emigrant to the United States and founder of an Albanian Orthodox church organization in Boston), favored land reform.

Zogolli's preference for consolidating central authority before pursuing nationalist claims to Kosovo and western Macedonia in the Yugoslav Kingdom earned him the wrath of Albanian ultranationalists, and his predilection for dictatorial methods sparked a powerful reformist-nationalist opposition movement, led by Noli and a Kosovar National Committee of emigrés. To allay public opinion while he dealt harshly with his opponents, Zogolli established a constitutional assembly and an interim government, assumed the premiership, and changed his name to Zogu because he felt Zogolli seemed too Turkish. His continued friendly dealings with the Yugoslav Kingdom intensified the ire of the Kosovar Albanian immigrants, who conducted guerilla actions against his authority and staged periodic local uprisings.

In early 1924, Noli led an uprising that forced Zogu to flee to the Yugoslav Kingdom. Noli's attempt at governing (1924), however, was foiled by Albania's social and linguistic divisions, the poor economy, no real state finances to speak of, the population's illiteracy, and rampant corruption. Political power lay with those who exerted the most brute force. While Noli talked of creating a modern, democratic, sound state, he lacked the means to achieve his goals. When lack of resources forced him to abandon Kosovo irredentism, he lost the nationalists' support, and his flirtation with obtaining the Soviet Union's aid so frightened everyone that he lost his domestic base. With Yugoslav assistance and promises to Italy and Britain of economic concessions, Zogu gathered an army in the Yugoslav Kingdom, invaded Albania (December 1924), and drove Noli out.

In 1925, Albania became a republic with Zogu as its first president. Despite outward constitutional trappings, he governed in dictatorial fashion. Realizing that the Albanian nationalists' irredentism regarding Kosovo made overt reliance on the Yugoslav Kingdom's support impractical, Zogu opened closer relations with Italy, which was eager to remain involved with Albania, constituted Albania's primary export market, and housed numerous Albanian emigrés. In 1927 Zogu and Mussolini signed an Albanian-Italian defensive alliance to counter a similar French-Yugoslav agreement. The treaty made Albania a veritable Italian protectorate. Zogu's government became dependent on Italian loans made in return for mineral concessions; Albania's military was placed under Italian control; and Italy was granted rights to build roads, bridges, port facilities, and schools in Albania. With Mussolini's approval, in 1928 Zogu was proclaimed King Zog I (1928-39) and given expanded authority.

With his head turned by the royal title, Zog steadily broke with his Italian benefactors. He rejected an Italian-proposed customs union (1932), tried some Italians in Albania for alleged anti-government plots, and closed Italian schools (1933). Zog was intimidated into reversing his anti-Italian policies by an Italian naval demonstration along the Adriatic coast (1934). There followed increased Italian control of Albania's military, expanded Italian trading and financial investments, and an influx of Italian colonists.

Not content with Zog, who continuously resisted Italian interference, Italy invaded Albania in the midst of the international furor surrounding Nazi Germany's takeover of Czechoslovakia (1939). Zog, whose dictatorial ways had alienated all elements in Albanian society, enjoyed scant support and was pushed out. Italian forces overran the state and a rump Albanian government was forced to accept Albania's union with Italy and recognize as ruler Italian king Victor Emmanuel III (1900-46). The Italian king, who never set foot in his new acquisition, exerted his authority through a Superior Fascist Corporative Council. In June 1939 Albania disappeared as an independent state.

YUGOSLAVIA
Dubrovnik
MONTENEGRO
Podgorica
Kotor
Peć
Priština
KOSOVO
Bajram Curri
Djakovica
Beli Drin R.
YUGOSLAVIA
Drin R.
Prizren
Kukës
Shkodër
Bar
Pukë
Tetovo
Skopje
Ulcinj
Drin i Zi R.
Lezhë
Peshkopi
ADRIATIC SEA
Krujë
Debar
Durrës
Tiranë
MACEDONIA
Prilep
Struga
Ohrid
Kavajë
Elbasan
Shkumbin R.
Bitola
Podgradec
Florina
Fier
Berat
Korçë
Kastoria
Çorovodë
Vlorë
Vijosë R.
GREECE
Tepelenë
Otranto
Gjirokastër
ITALY
Miles
0 10 20 30
Sarandë
0 10 20 30 40 50
Kilometers
CORFU
Ioannina
Albanian border, 1913-1921
Albania, 1921-1940
EPIROS

Map 42: The Balkans, 1939-1940

Certain socioeconomic and political interwar trends were noticeable in every Balkan state. Their mainly agrarian economies, supplemented by small industrial sectors, lay at the mercies of overpopulation and the international market; agrarian and socialist movements arose; and governments grew highly centralized and authoritarian. Those trends, combined with continuing national issues and the Great Depression's economic impact, drew the Balkan states into the Axis Powers' orbit and eventually into World War II.

Despite widespread land redistribution, Balkan agriculture remained at or a bit above subsistence level while rural overpopulation rose, creating permanent pools of poor, largely illiterate, and disgruntled "unemployed" rural workers susceptible to ultranationalist or socialist agitation. Limited industrial development and Communist or agrarian movements (like Stamboliiski's in Bulgaria) failed to overcome the rural problems. All of the Balkan states sought solutions in traditional autocratic political authority grounded in military power while maintaining a thin veneer of Western-style liberal democracy.

Nationalism remained a potent force in the interwar Balkans. Four semiofficial "Balkan Conferences" (Athens [1930], Istanbul [1931], Bucharest [1932], and Thessaloniki [1933]) did not overcome the states' burning national tensions. The Balkans' interwar situation somewhat resembled Western Europe's—similar national stresses led to similar dictatorial results in Germany and Italy. General European weariness after the past war's unparalleled devastation aided the rise to power of Benito Mussolini's Fascists (1922) in nationalistically unrequited Italy and Adolf Hitler's Nazis (National Socialists) in defeated Germany (1933). Those two revisionist states—wedded together as the Axis Powers (1936)—highlighted the Versailles order's weaknesses, demonstrating to the "loser" states that ultranationalist authoritarian (totalitarian) military dictatorships could break Versailles's restrictions.

The Balkan states paid close attention. Balkan nationalists of all stripes adopted fascist trappings—military dress, youthful uniformed paramilitary units, mass public demonstrations, authoritarian-tinged slogans and chants. The Croatian *Ustaše* and the Bulgaro-Macedonian IMRO established contacts with the Italian Fascists and German Nazis, while Romania spawned its own "Iron Guard" fascist movement. (See Map 37.)

In the wake of the Great Depression, which struck the Balkans in the early 1930s, the Axis Powers became the Balkan states' primary trading partners, purchasing Balkan cereals and tobacco products when no one else would. Italy monopolized Albania (see Map 41), and the rest of the Balkan market lay open to Germany. Eager to acquire cheap foodstuffs for his population and raw materials for his military, Hitler paid the Balkan states well and bought in large quantities through trade agreements tying payments to credits against purchases of German products. By 1939 Germany dominated the Balkan economies (except that of Albania), and when war erupted in that year, its economic position translated into political influence.

The kings of Bulgaria, Romania, and Greece ethnically were German and held pro-German sympathies. Bulgaria's King Boris had personal reasons for favoring the Axis—he was German by birth and married to the daughter of Italy's king Victor Emmanuel. With war imminent in 1939, the Ribbentrop-Molotov Pact partitioning Eastern Europe between Germany and the Soviet Union opened the possibility of Bulgaria aligning with Germany without offending its traditional Russian ally. Although not a fascist by nature or inclination, Boris drew closer to Nazi Germany, hoping to gain Macedonia and Dobrudzha. In the Treaty of Craiova (September 1940), Hitler forced his Romanian ally to return southern Dobrudzha to Bulgaria.

After Hitler's annihilation of Czechoslovakia (1939), Romania signed a trade agreement with Germany tying it closely to Hitler. The need to bail out a faltering economy overcame the fact that Hungary, Romania's national enemy, already was Hitler's satellite. Hitler needed both Hungary and Romania to dominate Eastern Europe. A faithful Hungary, itching to revise Trianon by force and kept at bay only by Hitler's power, ensured that its neighbors remained subservient to his will. Romania was a source of much-needed petroleum (which Germany lacked) and a potential source of manpower when Hitler initiated a planned invasion of the Soviet Union ("Operation Barbarossa").

Hitler forced Romania to relinquish Bessarabia and northern Bukovina to the Soviets (as stipulated in the Ribbentrop-Molotov Pact) and headed off an impending war over Transylvania between Hungary and Romania by giving the northern two-fifths of Transylvania to Hungary in the Second Vienna Award (August 1940). (See Map 38.)

Romanian nationalists found it difficult to accept the state's territorial losses. King Carol II was overthrown by a nationalist military coup (September 1940), and his son Mihai I (1940-47) was installed as successor. True power rested with Premier-Marshal Ion Antonescu, who was forced to accept the Craiova Treaty granting Bulgaria southern Dobrudzha and a German "protective" occupation of the Ploieşti oil fields (October) in return for Hitler's recognition of his dictatorship. As nationalist public outrage swelled, Antonescu expediently called on the Iron Guard, whose reign of terror on Jews and other political opponents brought Romania to the brink of economic collapse and political anarchy. Finally, Antonescu crushed the Guard by military force (January 1941) with Hitler's blessing. Hitler preferred a subservient Romanian dictator (whom he personally admired) to anarchistic fascists as he prepared to intervene against Greece before turning on the Soviet Union. (See Map 43.)

Danube R.
POLAND
USSR
SLOVAKIA
GERMANY
Vienna
Bratislava
BUKOVINA
AUSTRIA
Budapest
Iaşi
BESSARABIA
Kishinev
HUNGARY
Cluj
Ljubljana
YUGOSLAVIA
TRANSYLVANIA
Venice
Zagreb
Novi Sad
Timişoara
ROMANIA
Ploieşti
Belgrade
DOBRUDZHA
Zadar
Split
Sarajevo
Craiova
Bucharest
Niš
Vidin
Silistra
Danube R.
Varna
Podgorica
Priština
Dubrovnik
ITALY
BULGARIA
Sofia
Shkodër
Skopje
Plovdiv
Naples
Durrës
Tiranë
Ohrid
Edirne
Kavala
Istanbul
ALBANIA
Thessaloniki
Vlorë
GREECE
Ioannina
Larissa
MILES
0 50 100 150 200
0 100 200 300
KILOMETERS
TURKEY
Athens
Patras
Tripolis
DODECANESE IS.
(To Italy)
CRETE
Ribbentrop-Molotov Line, 1939
Romanian territorial losses to USSR, 1939
Second Vienna Award, 1940
Acquired by Bulgaria, Craiova Treaty, 1940
Occupied by Italy, 1939

Map 43: World War II – The 1941 Balkan Campaign

By keeping his Italian ally in the dark concerning his planned invasion of the Soviet Union and demanding that Italy make no move in the Balkans that might disrupt his secret preparations, Hitler's dispatch of troops to Romania in early October 1940 convinced Mussolini that Germany intended to acquire the Balkans for itself and leave Italy only Albania. Without informing Hitler, Mussolini decided to invade Greece from Albania in late October, win a spectacular victory (at a time when German military efforts apparently were frustrated after the Battle of Britain), and ensure a significant share of Balkan spoils. Hitler's annoyance with his ally's move rapidly turned to anger when the incompetently led Italian invasion forces were soundly trounced and routed by the Greeks.

Despite his growing prewar ties to Germany, Metaxas did not relinquish traditional Greek predilections for maintaining British support. Having blunted a simultaneous Italian assault against Egypt, Britain rushed troops and planes to mainland Greece and Crete and ships to the Adriatic. Assisted by those reinforcements, the Greeks counterattacked into Albania and the Italians barely managed to hold a tenuous defensive front in northern Epiros after suffering enormous losses in men and equipment. Desperate to patch together some way of directly bailing out his bumbling ally while still preserving his invasion preparations against the Soviet Union, Hitler forced Romania and Hungary to join the Axis alliance officially in late November. Bulgaria's March 1941 official entry into the Axis alliance permitted Hitler to amass German troops on Greece's northern border. In a final effort to position himself for a swift, overwhelming, and conclusive attack on Greece to end the diversion from his grand invasion plan caused by Italy, while simultaneously securing his southern flank for that operation, Hitler turned his attention to Yugoslavia.

In Yugoslavia, Serbian nationalists, already unhappy with Pavel's compromise with the Croatian nationalists, which gave the Croats an autonomous *Sporazum* within the state (see Map 36), were in no mood to accept the certain German military presence in Yugoslavia after Pavel, under intense pressure and the threat of force from Hitler, reluctantly joined Yugoslavia to the Axis alliance (March 1941). Nationalist anti-German army officers, aware that Britain had sent strong reinforcements to Greece, ousted Pavel (March), recognized King Petr II, and renounced the Axis pact. Infuriated, Hitler ordered an immediate assault on Yugoslavia and Greece. In an amazing display of extemporaneous military proficiency, his army general staff organized the invasion in a mere ten days. After a devastating terror bombing of Belgrade (6 April) paralyzed the Yugoslav high command, German forces, joined by Italian and Hungarian units, poured into Yugoslavia from Austria, Hungary, Romania, and Bulgaria. Overrun before its forces could be mobilized completely, Yugoslavia collapsed in 11 days and surrendered unconditionally (17 April). King Petr fled in exile to London. German losses totaled fewer than 600 men, while Yugoslavia suffered some 100,000 battle casualties and more than 300,000 men captured.

Greece also was invaded starting on 6 April. Pushing out of Bulgaria in the north, and boldly using armor over supposedly impassable terrain, German forces swiftly took Thessaloniki, forcing the surrender of cut-off Greek forces to its northeast, and drove back the British forces in central Greece. Their flank and rear threatened by Germans penetrating from Macedonia, the Greek forces facing the Italians retreated from Albania under constant pressure. Assaulted by the Germans from north, east, and south, by 23 April the Greek army's desperate resistance finally was overcome and Greece was forced to surrender. King George and Metaxas fled to London and established a government-in-exile.

The British troops fought their way through to the Peloponnese, from which they were evacuated to Crete (27 April) with the loss of all heavy equipment. A month later, a bloody German airborne assault (the first major such attack in history) conquered that island as well, capturing what remained of the British intervention forces. German casualties totaled some 10,000 men, while Greece suffered 70,000 killed and wounded and 270,000 captured, and Britain had more than 17,000 casualties and close to 12,000 men taken prisoner.

In under two months, Hitler became master of the entire Balkan Peninsula: Yugoslavia, Greece, and Albania were conquered outright; Bulgaria and Romania were allies. He dismantled conquered Yugoslavia into a number of territorial and ethnonational parts. Slovenia was divided between Germany and Italy. Vojvodina was shared between Hungary and local German residents. Most of Macedonia went to Bulgaria. The Italians acquired a slice of western Macedonia, Kosovo, Montenegro, most of Dalmatia, and portions of Bosnia-Hercegovina. Croatia Proper, Slavonia, most of Bosnia-Hercegovina, and the remainder of Dalmatia were turned over to an ultranationalist, neofascist Croatian puppet state ruled by Ante Pavelić, head of the *Ustaše*, who ostensibly served as the viceroy for an Italian absentee king. What was left of Serbia was placed under full and direct German military control, with Serb general Milan Nedić serving as the Serbian Marshal Pétain.

Although not dismembered like Yugoslavia, conquered Greece was forced to cede most of southern Epiros to Italian-dominated Albania (which also received Kosovo from its Italian masters), while Bulgaria acquired western Thrace and a contiguous section of Greek Macedonia, including the Aegean port-city of Kavala. Italy annexed the Ionian Islands in the Adriatic. The rest of Greece fell under direct German or Italian occupation, operating through a Greek puppet government in Athens.

Danube R.
GERMANY
SLOVAKIA
USSR
Bratislava
Vienna
AUSTRIA
II
Budapest
HUNGARY
BESSARABIA
Iaşi
Kishinev
Cluj
III
II
Ljubljana
Venice
Trieste
Rijeka
Zagreb
Timişoara
Novi Sad
PANZER
Bihać
YUGOSLAVIA
Ploieşti
Belgrade
ROMANIA
Zadar
Craiova
Bucharest
Split
Sarajevo
Danube R.
Silistra
Niš
Varna
Podgorica
Sofia
BULGARIA
Dubrovnik
ITALY
XII
Skopje
Plovdiv
Shkodër
Naples
Edirne
Durrës
Tiranë
Ohrid
Istanbul
XI
Thessaloniki
Vlorë
GREECE
Ioannina
Trikala
TURKEY
MILES
0 50 100 150 200
SICILY
Athens
Patras
0 100 200 300
Tripolis
Corinth
KILOMETERS
Axis Assault Operations and Army Concentrations
German
Hungarian
Italian
XII
III
XI
DODECANESE IS.
(To Italy)
Iraklion
CRETE

Map 44: The Axis-Dominated Balkans, 1941-1944

Following the 1941 Axis campaign, all of the conquered states were exploited rapaciously by their Axis masters. Arbitrary governance and police intimidation were the rule. The conquerors manipulated existing ethnonational rivalries and antagonisms to ensure their domination and turned a blind eye to the resulting pervasive atrocities. That especially was so in dismembered Yugoslavia, where such discord was diverse, deep-seated, and widespread.

The ethnonational situation within the divided state rapidly degenerated. Albanians, Hungarians, and Vojvodinan Germans indiscriminately massacred Serbs in the regions under their control. Bulgarians pressured Macedonian Slavs to adopt a Bulgarian ethnic identity. Pavelić's Croatia was the worst culprit. His *Ustaše* regime set out to either exterminate all Serbs and Jews in Croatia or forcibly convert them to Catholicism. A litany of massacres occurred and gruesome prison camps filled to overflowing. In Bosnia, the Muslims turned on the Serbs with a vengeance and joined in the bloodletting once political power shifted to the Croats. The situation in Bosnia came to resemble a veritable religious-cultural war, with Catholics and Muslims pitted against the Orthodox and Jews. The Serbs retaliated as best they could, and former Yugoslavia became a cultural battleground.

Axis-occupied Greece's wartime situation resembled that in occupied Yugoslavia. Axis satellite Bulgaria held parts of Greek Macedonia and Thrace; Italian puppet Albania controlled Greek Epiros; Italy occupied the bulk of mainland Greece and most Aegean islands; while Germany dominated strategically important Athens, Thessaloniki, Crete, and the crucial border with neutral Turkey. An Axis puppet government in Athens replaced that of King George II and Metaxas.

Occupied Greece suffered massive exploitation and repression. Hitler stripped the state of food and resources to supply his forces, and Britain's anti-Axis Mediterranean blockade included Greece. Since Greece's terrain greatly limited food production, and close to 50 percent of its population's food depended on imports, German policies and the British blockade brought on widespread famine. An estimated 7 percent of the Greek population and 30 percent of the state's national wealth were lost during the occupation. Food shortages, rampant inflation, and black-marketeering were the post-1941 wartime norm, and arrests, executions, and deportations (of Jews and slave laborers) were commonplace.

Although Hitler's allies Romania and Bulgaria suffered economic exploitation, the presence of German troops, and internal ethnonational problems, they at least preserved native self-rule and made national territorial gains. Romania, lying on the Soviet Union's border, could not avoid involvement in Hitler's assault on the Soviets (began 22 June 1941). In return for direct participation, Romania received Bessarabia and occupied additional territory to its east beyond the Dniester River (Transnistria). The fruits of Romania's pro-German involvement proved short-lived. In November 1942 the collapse of Romanian forces on the Eastern Front led to decisive German defeat in the Battle of Stalingrad, initiating a gradual Axis retreat. In 1943 the Ploieşti oil fields were destroyed by American and British air bombardment. By August 1944 the Soviets retook Transnistria and Bessarabia and lay ready to invade Romania.

The price Hitler paid for Bulgaria's participation in the 1941 Balkan Campaign was Macedonia. The second Bulgarian foray into the region proved even less happy than the first, in World War I. The Bulgarians comported themselves as occupiers and earned little gratitude from the natives for freeing them of Serbian control. Unlike Romania, King Boris refused an active role in Hitler's war against the Soviets, contenting himself with meeting his treaty obligations' minimum demands and attempting to consolidate control over Macedonia. Boris also somewhat undeservedly earned the distinction of saving Bulgaria's Jews from extermination in Nazi death camps. More through his minister Bogdan Filov's efforts and his wife's compassion than through Boris's own decision, trains loaded with approximately 15,000 Jews set to depart for the camps in 1943 never left Sofia. (Jews in Bulgarian-controlled Macedonia, however, were not saved.) Because Boris refused to declare war on the Soviets, Hitler summoned him to Berlin (1943). Soon after returning, he died under mysterious circumstances, and the leaderless Bulgarian government lapsed into ineffectiveness.

The turn of the Eastern Front's military tide against Hitler in 1943 sparked anti-Axis resistance movements throughout the Balkans. Though they shared a common general enemy, they were divided ideologically. Traditional nationalists (such as the Serbs led by the Hercegovinian Serb colonel Dragoljub [Draža] Mihailović and the Greeks under Colonel Napoleon Zervas) sought to restore prewar governing institutions and conditions. More radical movements—Communists, agrarians, and assorted socialists—espoused wide-ranging reforms or revolutionary changes. Radical anti-Axis guerillas, dominated by Communist partisans (particularly the Yugoslav Communist partisans led by Josip [Broz] Tito), were the most active and consistently effective. The anti-Axis Allies (Britain, the Soviet Union, and the United States), eager to assist those who inflicted the most damage on the enemy, and initially willing to postpone ideological problems until after the war, concentrated support on the radical guerillas (Tito's Communist partisans received more Allied military assistance than Mihailović's Serbian nationalist *Četniks)*.

The decision to support Communist partisans had repercussions for the postwar Balkans. In August 1944 Romanian king Mihai ousted Antonescu, ended military dictatorship, and joined the anti-Axis Allies. Romania's about-face opened the Balkans to the Soviet Red Army, which poured through Romania into Bulgaria and Yugoslavia. Axis resistance collapsed, and in its wake the various heavily armed Communist partisans, relying on immediate Soviet military support, swiftly gained control of the political situations on the ground everywhere except in Greece.

GERMANY
POLAND
USSR
(Partially to Germany)
SLOVAKIA
Danube R.
Bratislava
Vienna
AUSTRIA
Budapest
HUNGARY
Iași
Kishinev
Odessa
TRANSNISTRIA
BESSARABIA
Cluj
TRANSYLVANIA
Ljubljana
Trieste
Venice
SLOVENIA
ISTRIA
Rijeka
Zagreb
CROATIA
Timişoara
VOJVODINA
ROMANIA
BANAT
Belgrade
Ploieşti
Bucharest
Silistra
DOBRUDZHA
Danube R
Zadar
DALMATIA
BOSNIA-HERCEGOVINA
Sarajevo
SERBIA
Split
ITALY
MONTENEGRO
Podgorica
Dubrovnik
Rome
Sofia
Varna
BULGARIA
KOSOVO
Skopje
Plovdiv
Naples
Durrës
Tiranë
MACEDONIA
Edirne
ALBANIA
Ohrid
Kavala
THRACE
Istanbul
Vlorë
Thessaloniki
EPIROS
CORFU
GREECE
Ioannina
TURKEY
(Neutral)
MILES
0 50 100 150 200
0 100 200 300
KILOMETERS
SICILY
Athens
Patras
İzmir
DODECANESE IS.
(To Italy)
CRETE
ALBANIA States and regions annexed or occupied by Axis Powers, 1941:
by Germany
by Romania
by Italy
by Bulgaria
by Hungary

Era of Communist Domination
1944-1991

Map 45: Balkan Cominform States, 1945-1947

In August 1944, the Soviet Red Army entered the Balkans through Romania. Within days, Romanian king Mihai capitulated and Romania joined the Soviet war effort. Like dominoes, each Balkan Axis state fell before the onslaught. Communist partisan forces swiftly moved to gain political control while their overt partnership with the Red Army eased the way. By war's end in May 1945, every Balkan state except Greece (where British intervention prevented a Communist partisan takeover—see Map 46) lay under Communist control.

Romania joined the anti-Axis alliance (August 1944) and was occupied by the Red Army. Romanian Communists returning from exile in the Soviet Union were installed as the Soviet military's local officials, limiting King Mihai's actual governing authority. Direct Soviet intervention (February 1945) installed a "National Democratic Front" government, led by the pro-Communist agrarian Petru Groza (1945-52), and the Soviets systematically began stripping Romania of its wealth. America and Britain, hoping to preserve Allied unity with the Soviets, conditionally recognized the new government (February 1946) if it would hold open elections. When those elections were held (November 1946), Communist manipulation returned a "Front" majority. Romania then signed the Paris Treaty (February 1947), in which it lost Bessarabia and Bukovina to the Soviet Union in return for prewar Transylvania. The Communists eliminated all domestic opposition, King Mihai abdicated (December 1947), and the Groza government proclaimed Romania a "people's republic."

After Romania, Bulgaria lay open to Red Army assault. Desperate to avoid invasion, Bulgaria proclaimed its neutrality and abolished all fascist institutions. It was too late. The Soviet Union declared war on Bulgaria (5 September) and Bulgarian Communists, encouraged by the Red Army's proximity, staged a coup in Sofia (9 September) installing a Communist-led regency council. An armistice was declared and the Red Army occupied most of the state. Under the rubric of the "Fatherland Front," the Communists cemented their control, eliminating their Agrarian Union opposition through a Soviet-supported policy of terror. Elections (November 1945) returned a Communist puppet (Kimon Georgiev [1944-46]) as premier, but real authority lay with Communist leader Georgi Dimitrov (head of state, 1946-49), who orchestrated the total elimination of all remaining opposition. A rigged referendum abolished the monarchy (September 1946), after which a "people's republic" was proclaimed. Bulgaria signed the Paris Treaty, retaining only southern Dobrudzha of its wartime territorial spoils.

When the Red Army entered dismembered Yugoslavia (October 1944), it found actual military support from Josip Tito's Communist partisans. During the war, Tito embraced nationalist sentiments of all stripes to create a large, unified force, undermining Mihailović's Serbian nationalist *Četniks* by publicly proclaiming national toleration and self-determination ("Brotherhood and Unity") and tying the anti-fascist struggle to one creating a federal Yugoslavia. Tito planned and organized his future Communist Yugoslavia in wartime congresses of partisan representatives from every region of the former state. At a meeting in Jajce (1943), an alternative Yugoslav government under Tito (1943-80) was established, and the outlines of the planned state were finalized: Communist-controlled Yugoslavia was to be a federation of Serbian, Montenegrin, Croatian, Slovenian, Bosnia-Hercegovinian, and Macedonian republics, centrally controlled by the Communist Party from the capital at Belgrade. Kosovo and Vojvodina were recognized as autonomous Serbian provinces.

Soviet troops entered Serbia and, with Communist partisan assistance, took Belgrade (October 1944). Tito's partisans liberated the rest of Yugoslavia, exterminating all potential Croatian and Slovenian armed anti-Communist opposition in the process. Buoyed by authentic popular support as a war hero, Tito swiftly ended an initially thin liberal-democratic façade by calling for elections from a single-party ("People's Front") list (November 1945), after which the Yugoslav monarchy was abolished and a federal "people's republic" proclaimed. A new constitution (January 1946), modeled on the 1936 Soviet one, organized the state along lines outlined at Jajce. No pretense of liberal democracy was made—the Communists ruled the state and Tito ruled the Communists.

An offshoot of Tito's triumph was the Communist victory in Albania. From its wartime inception, the Albanian Communist Party, led by Enver Hoxha, essentially formed a branch of Tito's Yugoslav party, which held the actual power within Hoxha's "National Liberation Front," one of many Albanian antifascist resistance movements. With the Soviet invasion of the Balkans, civil war broke out among the various resistance groups, and Hoxha's Communists, aided by Tito, the Soviets, and Britain, emerged victorious, forming a countergovernment headed by Hoxha (1944-85) based on the Yugoslav model. They controlled most of the state by war's end, except for Kosovo, which Yugoslav partisans held for Yugoslavia. All serious opposition to Hoxha's regime was eliminated, and in elections demanded by the United States and Britain (December 1945), the Communists won in a landslide. The monarchy was abolished and Albania became a "people's republic" (January 1946), with a constitution giving the Communists undisputed political authority.

At the Yalta (February 1945) and Potsdam (July-August 1945) conferences of the victorious United States, Britain, and Soviet Union, Soviet leader Joseph Stalin (1924-53) was convinced that the Soviet Union required a buffer zone of "friendly" (Communist) states along its western border against the "capitalist" West. He upheld the Communist takeovers in Eastern Europe—creating the Soviet bloc and initiating the Cold War. Soviet control over its "friendly" states was cemented through the Cominform (Communist Information Bureau), founded to replace the former Comintern (Communist International), dismantled by Stalin during the war. Because of Tito's strident ideological loyalty, Cominform's headquarters were located in Belgrade.

POLAND
CZECHOSLOVAKIA
WEST GERMANY
Danube R.
Bratislava
Vienna
USSR
AUSTRIA
Budapest
BESSARABIA
Kishinev
Iaşi
SLOVENIA
HUNGARY
Cluj
Venice
Trieste
Ljubljana
Zagreb
ISTRIA
Rijeka
CROATIA
VOJVODINA
Timişoara
ROMANIA
YUGOSLAVIA
Novi Sad
Ploieşti
BOSNIA-HERCEGOVINA
Belgrade
Zadar
Craiova
Bucharest
Constanţa
Split
SERBIA
Sarajevo
Niš
Vidin
Danube R.
Silistra
MONTENEGRO
Tŭrnovo
Varna
Dubrovnik
Titograd
Priština
KOSOVO
BULGARIA
ITALY
Sofia
Burgas
Shkod r
Skopje
Plovdiv
Naples
Edirne
Durr s
Tiran
MACEDONIA
Ohrid
Kavala
Istanbul
ALBANIA
Thessaloniki
Vlor
Ioannina
Larissa
CORFU
GREECE
TURKEY
MILES
0 50 100 150 200
SICILY
Athens
Patras
0 100 200 300
KILOMETERS
Tripolis
RHODES
CRETE
Western and southern border of Cominform, 1945-1947
Reconstituted Yugoslavia, 1945
Autonomous provinces of Serbia

Map 46: The Greek Civil War, 1946-1949

Greece escaped Communist takeover at the end of World War II because of direct Western intervention. Thereafter, the Greeks consciously cultivated a self-image as "Western" Europeans by playing to classical Greek culture's historical importance for Western European civilization and by aligning themselves directly with the non-Communist West and joining NATO (1952) and the European Community/Union (EC/U, 1981). (See Map 47.) Despite those memberships, Greece persisted as a traditional Balkan state.

During World War II, Greek anti-Axis resistance movements spanned the political gamut from royalists to Communists. The most powerful was the National Liberation Front (EAM), which included Communists, socialists, agrarians, and a few traditional parties, whose military organization, the National Popular Party of Liberation (ELAS), was the strongest guerrilla force, controlling most of the state near the war's end. Politically left-leaning and intolerant of other Greek movements, EAM/ELAS recognized the government-in-exile in hope of playing an important postwar political role.

With British sponsorship, an interim coalition government was established under George Papandreou (1944-46), a staunch anti-Communist nationalist. Papandreou's government was installed in Athens (October 1944) under British protection and immediately decided to hold a referendum on the exiled king's return and to bring all armed resistance fighters under its authority. With Greece liberated, EAM/ELAS suspected Papandreou's motives for their disarmament, and a full-blown EAM/ELAS-royalist conflict erupted (December 1944). Only British military intervention saved the royalists from defeat. Stalin, in a pre-Yalta maneuver, warned EAM/ELAS to make accommodations with the British-backed authorities, so an agreement was reached (February 1945) calling for an interim regency and a plebiscite to determine the monarchy's fate, followed by parliamentary elections. Bolstered by British support, government ultrarightists then terrorized Communists and socialists, progressively driving EAM/ELAS partisans into Greece's north-central and Macedonian mountains.

EAM/ELAS refused to participate in British-demanded elections, so a rightist government was returned and King George II subsequently restored (September 1946). EAM/ELAS Communists, refusing to accept defeat, decided on civil war. A (Communist) Democratic Army commenced antigovernment military operations from its northern mountain strongholds, inflicting a string of defeats on government forces. Britain washed its hands of the deteriorating situation and handed it over to the United States (March 1947). President Harry S. Truman (1945-53), determined to combat communism's spread, dispatched military and economic aid to the Greek government (the initial act in the "Truman Doctrine"), bringing America into direct Balkan involvement for the first time.

Yugoslav Communist leader Tito became the Liberation Army's sponsor. The Communist rebels' strongholds along Yugoslavia's border facilitated their easy support. Proximity, however, proved a short-lived boon and a long-term liability. Massive Yugoslav Communist assistance made Tito and his Communist Macedonian Slav forces dominant in the Greek Communists' movement, and Macedonian Slavs came to constitute the majority in the rebels' ranks. The Greek Communists toed the traditional nationalist line regarding Macedonia, refusing to cede control over "Slavophone Greeks" to Yugoslavia's Macedonian Slav Communists and intending to retain Greece's territorial sovereignty. Internal tensions between the party's ethnic Greeks and Slavs intensified.

During the civil war's first year (1947), the Communist rebels' guerrilla tactics gained them most of mainland Greece. Their inability to capture a major town to serve as capital for their self-proclaimed provisional government, however, led to a change in tactics. Party leader Nikos Zakhariadis demanded that the army fight as a regular military force. The decision coincided with a transformation in the royalist government forces accomplished through American assistance and retraining. Changed rebel tactics and royalist air superiority turned the military tide against the Communists (late 1947).

The desperate Communists resorted to forced general conscription in the Macedonian regions under their control and evacuated children to Soviet-bloc states hoping to strengthen their position. Instead, those actions backfired, eroding their popular support and increasing the ethnic Slavs' numbers in their ranks—heightening internal Greek-Slav tensions.

In 1948 the Tito-Stalin split (see Map 47) doomed the Greek Communist movement to failure. The break immediately strained relations between Yugoslav and Greek Communists and divided the Greek Communist leadership. The majority, convinced that the Soviet Union was global communism's bastion, sided with Moscow. Unfortunately for their cause, they faced a Faustian choice of embracing either Tito or Stalin. Preoccupied with defending his position against the Soviets and their Balkan allies and maintaining control over all of Yugoslavia, including Yugoslav Macedonia, Tito grew unwilling to assist the Greek Communists. Stalin, hoping to undermine Tito's position within Yugoslavia, and convinced that the Greek Communists' cause was not worth a direct confrontation with the United States, endorsed the idea of creating an independent Communist Macedonian state that would include Greek Macedonia. If the Greek Communists retained their ties with Tito, they were doomed to defeat through lack of support. If they sided with the Soviets, they would be forced to accept Greece's future dismemberment.

The Greek Communists decided against continued reliance on Tito (early 1949). In July Tito closed his border with Greece, after which, alone and poorly equipped, they faced the modernized Greek army commanded by General Alexander Papagos. Papagos defeated the Communist forces in northwestern Greece and drove the remnants of the Democratic Army into Albania (late Summer 1949). The defeated Greek Communist leaders declared a temporary cease-fire (October). Although their exiled fighters were kept on a war footing for some years, the civil war ended.

YUGOSLAVIA
Skopje
Vardar R.
Plovdiv
BULGARIA
Edirne
Tiranë
MACEDONIA
Serres
Kavala
THRACE
Istanbul
ALBANIA
Florina
Naoussa
Thessaloniki
Konitza
Metsovon
EPIROS
Trikala
Larissa
GREECE
Ioannina
THESSALY
Arta
Karditsa
Karpenision
Lamia
LESBOS
TURKEY
İzmir
Patras
Athens
SAMOS
Pirgos
Trikala
IKARIA
RHODES
Miles
0
50
100
Iraklion
CRETE
0
50
100
150
Kilometers
Territories controlled by Communist guerillas, late 1948
Final stronghold of Communist guerillas, late 1949

Map 47: Splits in Communism, 1948-1960

If the Soviets' postwar creation of a Communist state buffer zone in Eastern Europe brought on the Cold War, the first strategic action in that conflict was the United States' Marshall Plan (1948), offering economic assistance to European states in return for acceptance of liberal democracy, capitalism, American market penetration, and an American military alliance. The Soviets countered (1949) with the Council of Mutual Economic Assistance (Comecon) for its European satellites. The Berlin Blockade (1948-49) and Communist victory in China (1949) convinced the United States and its allies that Western military unity was needed, resulting in the North Atlantic Treaty Organization (NATO, 1949). In response, the Soviets formed a similar organization, the Warsaw Pact (1955), until which Cominform was the heart of the Soviet "bloc," ensuring conformity with Soviet ideological, political, economic, and social models. Throughout the Cold War era, however, Communist Yugoslavia, Albania, and Romania markedly diverged from official Soviet positions.

Flattered by Stalin's postwar favor, and proud of his partisan record, Tito's somewhat inflated self-image led to misunderstandings with the Soviets. Convinced that Stalin wished him to lead an active anti-Western, revolutionary foreign policy offensive, Tito attempted to found a Communist Balkan federation and increased assistance to the Greek Communists in the Greek Civil War (see Map 46) without consulting Stalin. Stalin, fearing that Tito would undertake adventurous foreign initiatives counter to Soviet interests, demanded that Tito seek his permission beforehand. Tito, convinced that Stalin was compromising ideology for the Soviets' own sake, refused. Stalin then convened a Cominform meeting in Bucharest (June 1948), which condemned Tito and called on the Yugoslav Communists to change either their policies or their government. Tito denounced the Soviets.

Soviet reaction to Tito's defiance was swift. The Cominform states severed relations with Yugoslavia and purged all party members suspected of associations with Tito (as well as all opposition elements, Communist or not). Incidents erupted along Yugoslavia's borders with neighboring Cominform states and border fortifications were erected. The Yugoslav Communists were attacked in the media, support was given to anti-Tito Yugoslav opposition groups operating abroad, and plans for Yugoslavia's dismemberment into independent Communist states were floated.

Isolated within the Communist world, Tito had no option but to turn to the Western Powers that he previously had reviled. While refusing to compromise Marxist-Leninist tenets, he pragmatically sought Western assistance and opened diplomatic contacts with Britain and the United States. By the 1960s the United States, once a hated ideological enemy, was one of Communist Yugoslavia's largest sources of foreign aid, and Yugoslavia was dependent on non-Communist Western military and economic support.

Tito strengthened his domestic position by attempting to demonstrate that a state could embrace communism without relinquishing its unique national identity and independence. The Yugoslav Communists reconsidered their guiding socialist principles and concluded that Stalin was the actual "deviationist" while they were the "true" Marxists-Leninists, producing a counter-Soviet doctrine (January 1950). Worker self-management and decentralized economic planning ultimately led to a "mixed" socialist-capitalist system, in which the Communists retained dictatorial political control but no longer were inseparable from the economic structure. Although Yugoslav-Soviet relations improved during the "de-Stalinization" process initiated by Soviet leader Nikita Khrushchev (1953-64), Communist Yugoslavia remained independent of Soviet control.

Enver Hoxha's post-World War II Communist Albania initially existed as a satellite of Tito's Yugoslavia, which threatened Albania's continued state sovereignty. Aware of Albanian ethnic uniqueness and the Slavs' sorry record regarding Albanian civil and political rights in former Yugoslav nation-states, the Albanian Communists feared that foreign Italian domination would be replaced by equally foreign Slavic mastery. They joined the anti-Tito offensive, broke completely with Yugoslavia, and adopted strict Stalinist authoritarian political and economic models. With the post-Stalin Yugoslav-Soviet rapprochement, the Albanian Communists grew concerned about their continued independence, fearing that Khrushchev might buy Yugoslav support at Albania's expense. Apprehension grew as Soviet de-Stalinization progressed. Albania publicly assumed an anti-Soviet stance and cast about for an alternative ally. In 1960 it found one in Maoist China. The adoption of Maoism led to Albania's near total international isolation and severely retarded economic development.

In Romania, Gheorghe Gheorghiu-Dej (1953-65) came to power when the "Front" government was dissolved (1953). By the end of the 1950s Romanian society was totally Sovietized. Using the 1956 crises in Poland and Hungary as pretexts, Gheorghiu-Dej reasserted Romania's national independence by moving toward a foreign policy semi-independent of the Soviets. Romania developed its own brand of "nationalist communism," strengthening Communist authoritarian control by officially embracing traditional anti-Russian and anti-Hungarian Romanian nationalism. Gheorghiu-Dej's policies were continued and intensified by Nicolae Ceauşescu (1965-89), who parlayed his position into a veritable Phanariote-like personal, feudal-style autocracy. He aggravated tensions with Hungary with his policies against the Transylvanian Hungarian minority and bombastically voiced Romanian nationalist claims to Soviet Bessarabia and Bukovina. His anti-Soviet antics in the late 1960s won Romania tentative "Most Favored Nation" status with the United States (1970s), but his domestic policies, creating a nepotistic and dependent government and party system granting him overwhelming dictatorial authority, virtually returned Romania to Ottoman-era sociopolitical conditions.

WEST GERMANY
CZECHOSLOVAKIA
U S S R
Danube R.
Bratislava
Vienna
AUSTRIA
Budapest
BESSARABIA
HUNGARY
Iaşi
SLOVENIA
Cluj
Venice
Ljubljana
Zagreb
Trieste
Rijeka
CROATIA
VOJVODINA
ROMANIA
BOSNIA-HERCEGOVINA
Sarajevo
Novi Sad
Zadar
Belgrade
Bucharest
SERBIA
Constanţa
Split
Danube R
Silistra
ITALY
YUGOSLAVIA
MONTE-NEGRO
Niş
Dubrovnik
KOSOVO
Varna
Sofia
Rome
Titograd
Priština
BULGARIA
ALBANIA
Skopje
Naples
Plovdiv
Tiranë
MACEDONIA
Edirne
Istanbul
Thessaloniki
Ioannina
GREECE
TURKEY
SICILY
Athens
MILES
0 50 100 150 200
0 100 200 300
KILOMETERS
RHODES
Warsaw Pact states, post 1960
GREECE NATO states, 1949-1991
CRETE

Map 48: Collapse of Communism, 1989-1991

Communists portrayed their success in acquiring state power as the class struggle's inevitably victorious outcome, leading to a totally egalitarian workers' society. To construct that objective, the Communists asserted that their "temporary" dictatorial authority was necessary to ensure "proper" development through central economic planning.

By the 1980s, Soviet-style communism's faults were manifest. Despite industrial nationalization and agricultural collectivization, central planners could not control or account for every economic factor, and attempts to micromanage those that they recognized produced a continually expanding, dysfunctional bureaucracy that crippled economic productivity and development. Inherent consumer shortages, lack of bureaucratic coordination, and administrative irresponsibility produced economic stagnation and demoralization. Cold War-era détente exposed Communist societies to Western capitalist materialism and media with negative comparative consequences. While the West's Cold War strategy of increased military spending may have hastened communism's demise, the Communist system's innate flaws doomed it to internal collapse.

That collapse began with Soviet leader Mihail Gorbachev's (1985-92) reform efforts. His *perestroika* (restructuring) and *glasnost* (openness) policies were intended to address the Communist system's economic planning and political authoritarian deficiencies. They were copied, in some form or another, by most Warsaw Pact states. The results, however, bore little relation to the intent. The original perestroika program failed and caused economic chaos. Glasnost grew from promoting openness into the Communist Party's complete reorganization, reducing it to a secondary governmental role. Gorbachev became caught in a dilemma when the openness that he fostered resulted in the reemergence of nationalism among the Soviet Union's diverse population. Only Communist government force could prevent the Soviet system's outright collapse, but its use would discredit him and destroy his reforms. Thus, he presided over the demise of Soviet communism and the Soviet Union itself.

With the Soviet Union in turmoil, Gorbachev notified the Warsaw Pact governments (1988) that they were on their own in facing their economic problems and the resurgence of nationalism. By that time, all of the Balkan Communist states were crippled economically and suffering internal nationalist tensions.

Bulgaria followed Gorbachev's reform lead with more lip service than implementation. By 1989 authentic economic reform desperately was needed since Bulgaria's moribund economy was on the verge of expiring. The Communist government of Todor Zhivkov (1954-89) tried to prop up sagging popular support by playing the Bulgarian nationalist card against the Turkish Muslim minority, forcing on them name changes and enacting anti-Muslim cultural policies (1985-86). Muslim unrest built and, as Soviet communism unraveled in 1989, a desperate Zhivkov initiated a "Bulgarian unity" campaign that forced some 360,000 Turks and Muslims to flee to Turkey, later sparking mass demonstrations of Bulgarian dissidents that eventually succeeded in removing Zhivkov. Although the Communists tried to retain political control, by 1991 they were ousted democratically and Bulgaria began a long transition toward post-Communist rebuilding.

By 1989, conditions in Romania were miserable. There were violent street demonstrations in Bucharest and some Transylvanian cities; relations with Hungary were bad; the economy was crippled; and drastic rationing, brought on by Ceauşescu's efforts to pay off Romania's foreign debt, demoralized the population. Although reformers and dissidents founded an underground National Salvation Front (NSF), the security forces kept it in check. In December 1989, Ceauşescu's security police tried to deport the dissident Transylvanian Hungarian pastor László Tökés but met strong resistance from the Timişoara Hungarians, which escalated into rioting and calls for Ceauşescu's demise. At a staged public rally in Bucharest to condemn the demonstrators, Ceauşescu was greeted with catcalls, and his contemptuous reaction sparked antigovernment demonstrations that, fed by years of misery and repression, turned into a bloody revolution led by the NSF. Ceauşescu and his wife were captured, summarily tried, and executed (on live television, broadcast internationally) by enraged rebels on Christmas Day (25 December 1989). The "Christmas Revolution" overthrew the Communist Party, but the new NSF interim government embraced the authoritarian style of Communist rule, causing Romania much political turmoil long after communism ended.

By 1989, Yugoslavia's economic woes had reached staggering proportions. Inflation ran at 300 percent; financial irresponsibility was rampant; domestic and foreign debt was crushing; and all economic indicators were in decline. Partly because of Tito's legacy, and partly out of pride over their record of independence, the Yugoslav Communists entered the Gorbachev-inspired reform era haltingly. The few reforms attempted by federal authorities were stymied by the republics' failure to ratify them. Following the rise in Serbia of Communist strongman Slobodan Milošević (1987-2000), reforming communism became less of a priority in multinational Yugoslavia than embracing resurgent nationalism. Ultimately, nationalism destroyed the Tito-created Yugoslav state and initiated a decade of national warfare. (See Maps 49 and 50.)

Albania's was the last Balkan Communist regime to fall. The Communists, headed by Ramiz Alia (1985-92), paid scant attention to the Gorbachev-era reform movement until Ceauşescu's fall. A series of dissident demonstrations (1989-90), coupled with the example of Ceauşescu's fate, awakened Alia to the danger facing his regime. He issued limited reform measures (1990), but the population, their condition nearly as miserable as the Romanians', demanded democratic measures. Alia was forced to accept the formation of non-Communist political parties, after which the opposition succeeded in bringing down the Communist government (June 1991). An opposition multiparty coalition government of "national stability" was installed and communism ended.

POLAND
CZECHOSLOVAKIA
GERMANY
USSR
Danube R.
Bratislava
Vienna
AUSTRIA
Budapest
HUNGARY
Iași
BESSARABIA
Kishinev
Trieste
SLOVENIA
Venice
Ljubljana
Zagreb
Cluj
Rijeka
CROATIA
Timişoara
Vukovar
Bihać
Novi Sad
ROMANIA
BOSNIA-HERCEGOVINA
YUGOSLAVIA
Belgrade
Bucharest
Zadar
Knin
Sarajevo
SERBIA
Constanţa
Mostar
Niš
Danube R.
Silistra
MONTENEGRO
Dubrovnik
Podgorica
Priština
Tŭrnovo
Shumen
Varna
KOSOVO
ITALY
Prizren
Sofia
BULGARIA
Plovdiv
Burgas
Shkodër
Skopje
Naples
Durrës
Tiranë
MACEDONIA
Edirne
ALBANIA
Ohrid
Kavala
THRACE
Istanbul
Thessaloniki
Vlorë
Ioannina
Larissa
GREECE
TURKEY
MILES
0 50 100 150 200
0 100 200 300
KILOMETERS
SICILY
Athens
Patras
Tripolis
RHODES
CRETE
Yugoslav successor states, 1991
Areas of active ethnonational unrest, 1980-1991
Area of national revolution, 1989

Era of Post-Communism
1991-Present

Map 49: Wars of Yugoslav Succession, 1991-1995

Communism ended in Yugoslavia amid a welter of resurgent nationalism. The post-World War II Yugoslavia was the federal construct of Tito, whose consolidation of power in the state depended on embracing the assorted national aspirations of prewar Yugoslavia's ethnic populations. Only Tito's iron will to maintain his central authority and determination to assert the overriding priority of Communist ideology kept the state together. His split with Stalin (see Map 47) led to the formation of Yugoslav "national communism," with the emphasis on "Yugoslav" (rather than on Yugoslavia's various constituent nationalities).

Titoist federalism was intended to satisfy the prewar nationalist ambitions of Yugoslavia's diverse population within the overall context of a unitary Communist state by providing six Yugoslav "nationalities" with their own republics. While the Serbian, Montenegrin, Croatian, and Slovenian republics were predicated on long-standing national identities, Macedonia appeared as an official national entity for the first time in its history (with a specifically manufactured ethnic identity), and Bosnia-Hercegovina, ethnically and religiously divided, was recognized as a unitary republic. Despite imposed federalism, however, Communist Yugoslavia was the most volatile focus of latent nationalist strife in the Communist-era Balkans.

Although federalism initially satisfied many prewar Croatian nationalist claims, Yugoslav communism's decentralization increasingly loosed restraints on intellectual activities (among which were resurrected allusions to prewar nationalism). Intense debates between Croats and Serbs over the nature of "Yugoslavism" degenerated into old national sentiments and mutual distrust by the early 1970s. For their part, the Serbs perceived national problems stemming from Serbia's autonomous, mostly non-Serb provinces of Kosovo and Vojvodina. They had been granted voting privileges equal to the republics' (Serbia's) in the central federal government (1966) and non-Serbs came to dominate their administrations. Regarding the Muslim majority in Bosnia-Hercegovina, the federal government never could decide officially if or how they constituted an "ethnic nationality" (as opposed to being either Croats or Serbs), while the Bosnian Muslims acquired a political structure that resembled more a *millet* than an ethnonational identity (giving them alone an awareness of Bosnia-Hercegovina as a unique political entity). When a revolving collective federal presidency among leaders of the constituent republics was instituted following Tito's death (1980), growing national tensions led to its general ineffectiveness.

By the 1980s, Serb-Albanian nationalist tensions in the Serbian autonomous province of Kosovo broke out into sporadic open conflicts (see Map 50), opening the way for Serbian Communist strongman Slobodan Milošević to play the nationalist card in his bid for authoritarian power at the time that the Gorbachev reforms' failures were unraveling European communism. (See Map 48.) Beginning in 1988 Milošević championed the tenets of "Greater Serbia" nationalism within Yugoslavia, trampling on the autonomy of the Kosovo and Vojvodina Serbian provinces and raising the specter of Serbian domination throughout the federal state. The Croats, Slovenes, and Macedonians, resentful of de facto Serbian predominance prior to Milošević, found the new situation unacceptable and initiated a series of political moves to disassociate themselves from Serbian control. As tensions mounted, Milošević supported Serbian ultranationalists in Croatia Proper, Slavonia, and Bosnia-Hercegovina, which ultimately brought about the succession of Slovenia and Croatia from Yugoslavia (June 1991) as independent, non-Communist national states. By year's end Bosnia-Hercegovina and Macedonia followed their lead.

Milošević sent the Yugoslav National Army (JNA) into Slovenia (June 1991) but it was stymied humiliatingly, after which it set upon Croatia in support of Serbian nationalists who rose in eastern Slavonia and Croatia Proper, where they had declared an independent region called Krajina centered on Knin. Vicious fighting between Serbs and Croats erupted around Vukovar, and the first evidence of the Serbian "ethnic cleansing" policy emerged. Despite Western European recognition, Croatia, led by the nationalist Franjo Tudjman (1991-99), was defeated and lost a quarter of its national territory.

Fighting then spilled over into Bosnia-Hercegovina, whose Muslim leader Alija Izetbegović (from 1991) unsuccessfully attempted to maintain a united state in the face of Milošević-supported Serbian nationalists led by Radovan Karadžić, who proclaimed a Serbian Republic of Bosnia-Hercegovina (March 1992) in unity with Milošević-led Serbia. In April 1992, fighting erupted between the two sides in Sarajevo and swiftly spread throughout the state. The Bosnian Serbs placed Sarajevo under siege and, with help from Milošević's JNA, gained control of two-thirds of the state by the end of 1992. A concerted "ethnic cleansing" policy was conducted in territories under Serbian occupation in an attempt to cement a permanent Serbian ethnic presence. Croatian nationalists joined in the fighting against the Muslims. Those in Hercegovina set up a Croat Herceg-Bosna republic (July 1992) and initiated their own "ethnic cleansing" activities.

The three-way war raged on through 1993 until 1995, resulting in enormous casualties, mass emigrations, and widespread destruction of private and historical property (including the historic Ottoman-era bridge in Mostar). Failed EC/U peace efforts led to a string of United Nations (UN) attempts with similar results. UN peacekeeping forces proved ineffectual in stemming the fighting or the atrocities, but a Muslim-Croat federative alliance (1994) helped stabilize the anti-Serb military position. In May 1995 Croatia, rearmed with United States support, launched a successful attack against Serbian positions in Croatia. Threatened with military collapse and NATO air intervention, and pressured by Russian diplomacy, Milošević reined in his Bosnian Serb surrogates and signed an American-brokered peace accord for the Bosnian war in Dayton, Ohio (November 1995), tentatively preserving a Muslim-Croat and Bosnian Serb federated state of Bosnia-Hercegovina.

HUNGARY
SLOVENIA
Danube R.
Varaždin
Ljubljana
Zagreb
Drava R.
CROATIA
VOJVODINA
Osijek
SLAVONIA
Novi Sad
Rijeka
Glina
Okučani
Vukovar
Sava R.
KRAJINA
Bihać
Banja Luca
Brčko
Belgrade
SERBIAN
REPUBLIC
Tuzla
Maglaj
SERBIA
Jajce
BOSNIAN GOVT.
Knin
BOSNIA
Srebrenica
Zadar
Žepa
DALMATIA
HERCEG-BOSNA
Sarajevo
Pale
Goražde
YUGOSLAVIA
Split
Mostar
Miles
0 20 40
0 20 40 60
Kilometers
MONTENEGRO
KOSOVO
CONQUESTS IN SLAVONIA
AND BOSNIA, 1991-1994
Dubrovnik
Podgorica
Bosnian border
Croatian holdings
Bosnian (Muslim) govt. holdings
Serbian holdings
ALBANIA

HUNGARY
SLOVENIA
Danube R.
Varaždin
Ljubljana
Zagreb
Drava R.
CROATIA
VOJVODINA
Osijek
SLAVONIA
Novi Sad
Rijeka
Okučani
Vukovar
Glina
Sava R.
Bihać
Banja Luca
Brčko
Belgrade
Tuzla
SERBIA
Jajce
Maglaj
Knin
Srebrenica
BOSNIA
Žepa
Zadar
Sarajevo
Pale
DALMATIA
Goražde
YUGOSLAVIA
Split
Miles
0 20 40
Mostar
0 20 40 60
Kilometers
MONTENEGRO
KOSOVO
THE "DAYTON ACCORD"
SETTLEMENT, 1995
Dubrovnik
Podgorica
Bosnian border
Muslim-Croat Federation
Serbian Republic
U. N. presence in eastern Slavonia
ALBANIA

Map 50: The Kosovo Crisis, 1999

In 1966 Tito granted Serbia's autonomous provinces equal voting privileges with republics at the federal level, resulting in greater Albanian participation in Kosovo's provincial administration. Until that time, the province's Albanian majority (nearly 90 percent of the inhabitants) had lived under Serbian administrative dominance and experienced persistent discrimination from the Serb authorities. The new Communist Albanian provincial representatives tended to act in a retaliatory fashion toward the resident Serbian minority, and also began calling for Kosovo's elevation to republic status. The post-Tito Yugoslav collective federal presidency refused to accede to the Kosovar Albanians' desires, since doing so would have violated Communist Yugoslavia's constitutional foundations. (Constitutionally, only Yugoslav "nationalities" could possess their own republics; since an Albanian nation-state already existed outside of Yugoslavia, the Kosovar Albanians technically were a Yugoslav "national minority," despite their provincial majority, so were disqualified from having a republic.)

Albanian nationalist ideals took hold among university students, disaffected intellectuals, and workers throughout Kosovo, and in 1981 they staged demonstrations and riots. Eventually, a third of the JNA was deployed in Kosovo to put a lid on the disturbances. In the riots' aftermath, a national paranoia emerged among the Kosovar Serb minority. Exaggerated reports of Albanian anti-Serb atrocities appeared in the Serbian media throughout the 1980s, fueling a growing Serbian nationalist xenophobia. The paranoia in Serbia over Kosovo exploded in 1989 during a Serbian nationalist celebration of the 600th anniversary of the medieval battle at Kosovo Polje. Slobodan Milošević's public expression of Serbia's support for the Kosovar Serbian nationalists at that event reopened old issues stemming from Yugoslavia's nature as a unitary federation of separate, distinct "nations" that essentially was dominated by Serbs, and was the spark that ignited Yugoslavia's disintegration. (See Map 49.)

Serbian leader Milošević reduced the autonomy of Serbia's Kosovo and Vojvodina provinces (1989), and units of the JNA were stationed in Kosovo to quash increasing Albanian nationalist riots and demonstrations, which occurred throughout 1990. Many leading Albanian intellectuals, officials, and administrators were arrested; the Serbian constitution was amended to eradicate the remaining vestiges of Kosovo's autonomy (1990); and new laws were passed attacking the Kosovar Albanians' civil rights, language, and culture. The Kosovar Albanians continued their calls for Kosovo's elevation to republic status and Serbian anti-Albanian repression intensified—Albanian property rights were restricted, and new employment laws expelled more than 80,000 Albanians from their jobs.

In September 1990 a group of Kosovar Albanian nationalists proclaimed an independent Kosovo republic. Within a year (September 1991) they conducted a referendum among fellow disaffected Albanians that won popular support for their program but increased the ire of the Serbian authorities. By the end of 1991, Yugoslavia dissolved into five separate states—Slovenia, Croatia, Bosnia-Hercegovina, Macedonia, and a truncated Yugoslavia consisting of Serbia and Montenegro. Amid the shock of the initial Serbo-Croat war in late 1991, Albania extended unilateral recognition to an independent sovereign Kosovo in support of the repressed Kosovar Albanian nationalists. In May 1992, a secret election was held throughout Kosovo to create a Kosovo republican government, and Ibrahim Rugova, leader of a political party of intellectuals, was elected president. Rugova's Gandhi-like political approach primarily emphasized the rejection of Serbia's continued governing legitimacy in Kosovo, nonviolent opposition to the Serbian authorities, and winning international support for the Kosovar Albanians' national cause.

Throughout the years of warfare in Bosnia (see Map 49), the Kosovar Albanians' situation deteriorated. Albanian land was confiscated and handed over to Bosnian Serb refugees, who were settled as Serb colonists in Kosovo. Increased numbers of Albanians were thrown out of work, expelled from the university, or arrested, gradually discrediting Rugova's government among the general Albanian population. His ineffectiveness was highlighted when NATO and EC/U authorities essentially (but mistakenly) embraced Milošević as the guarantor of Balkan peace at the Dayton meeting (late 1995) ending the Bosnian debacle.

Throughout 1996 and 1997, a small group of Kosovar Albanian nationalists broke with Rugova and created the guerilla Kosovë Liberation Army (KLA). They instigated an uprising in 1998 that resulted in intensive Serbian military operations and anti-Albanian atrocities in Kosovo. Bolstered by general Serbian nationalist sentiment, Milošević steadfastly refused to temper Serbian treatment of Kosovar Albanians in the face of growing expressions of international displeasure. By late 1998, JNA forces had the KLA on the run, Serbian ultranationalist police and paramilitaries stepped up their anti-Albanian campaign of atrocities, and thousands of Kosovar Albanians fled the province in terror, while Milošević played cat-and-mouse diplomatic games with United States and other Western representatives seeking to bring about an end to the violence.

As the JNA buildup in Kosovo continued in early 1999 and atrocity stories multiplied, a last-ditch effort to forge a solution to the situation was made at Western insistence in Rambouillet, France, but broke down over mutual Serbian and Kosovar Albanian intransigence. In March, the Serbs unleashed their military against the KLA, and an orgy of "ethnic cleansing" was initiated. Over a million Albanian refugees flooded into neighboring Albania and Macedonia. Led by the United States, NATO air forces pounded Serbia for over two months before Milošević conceded defeat and withdrew his forces from the province (June). NATO peacekeeping troops moved into Kosovo to protect the returning refugees and prevent future violence. Despite the NATO presence, Albanian anti-Serb reverse "ethnic cleansing" persisted, and no definitive solution to the province's national problems was reached.

YUGOSLAVIA
SERBIA
MONTENEGRO
ALBANIA
MACEDONIA
Raška
Niš
Novi Pazar
Kuršumilja
Leposavić
Leskovac
Mitrovica
Podujevo
Rošaj
Peć
Plav
Priština
Kosovo Polje
Gračanica
Novo Brdo
Kamenica
Vranja
Dečani
Gnjilane
Dakovica
Bajram Curri
Preševo
Kačanik
Prizren
Kukës
Blače
Kumanovo
Skopje
Tetevo
Gostivar
Veles
Ibar R.
Morava R.
Sitnica R.
Beli Drin R.
Drin R.
Drin i Zi R.
Vardar R.
I
II
III
IV
V
Border of Kosovo province, Serbia
NATO peacekeeping sector boundaries
NATO peacekeeping sectors
I French
II Italian
III British
IV German
V American
MILES
0
10
20
0
10
20
KILOMETERS

Selected Bibliography

The following list of sources is not comprehensive. For the most part, the works are general studies or commonly available atlases. Some have been included because their usefulness is broader than any single topic or because it is believed that they are of intrinsic interest to the general reader or student. The envisioned audience for this concise atlas is assumed to be English-speaking general readers and students, thus only works written in English are listed. While this might seem arbitrarily limited in scope, it is not. Those who possess the ability to read non-English languages will find plenty of additional titles to explore in the bibliographies and notes to many of the works cited.

Anzulovic, Branimir. *Heavenly Serbia: From Myth to Genocide.* New York: New York University Press, 1999.

Babinger, Franz. *Mehmed the Conqueror and His Time.* Translated by Ralph Manheim. Princeton, NJ: Princeton University Press, 1978.

Banac, Ivo. *The National Question in Yugoslavia: Origins, History, Politics.* Ithaca, NY: Cornell University Press, 1993.

Barker, Elisabeth. *Macedonia: Its Place in Balkan Power Politics.* London: Royal Institute of International Affairs, 1950.

Barraclough, Geoffrey, ed. *The Times Atlas of World History.* Revised ed. London: Times Books, 1979.

Benderly, Jill, and Evan Kraft, eds. *Independent Slovenia: Origins, Movements, Prospects.* New York: St. Martin's Press, 1994.

Brown, L. Carl. *The Imperial Legacy: The Ottoman Impact on the Balkans and the Middle East.* Boulder, CO: East European Monographs, 1995.

Cambridge Medieval History. Vol 1: Map supplement. Cambridge: Cambridge University Press, 1911.

Carter, Francis W. *An Historical Geography of the Balkans.* London: Academic Press, 1977.

Castellan, Georges. *History of the Balkans from Mohammed the Conqueror to Stalin.* Translated by Nicholas Bradley. Boulder, CO: East European Monographs, 1991.

———. *A History of the Romanians.* Translated by Nicholas Bradley. Boulder, CO: East European Monographs, 1989

Clogg, Richard. *A Concise History of Greece.* Cambridge: Cambridge University Press, 1992.

Crampton, Richard J. *Bulgaria, 1878-1918: A History.* Boulder, CO: East European Monographs, 1983.

———. *A Concise History of Bulgaria.* Cambridge: Cambridge University Press, 1997.

———. *Eastern Europe in the Twentieth Century—And After.* 2nd ed. London: Routledge, 1997.

Crampton, Richard J., and Ben Crampton. *Atlas of Eastern Europe in the Twentieth Century.* London: Routledge, 1996.

Darby, H. C., and Harold Fullard. *The New Cambridge Modern History.* Vol. 14: *Atlas.* Cambridge: Cambridge University Press, 1978.

Donia, Robert J., and John V. A. Fine, Jr. *Bosnia and Hercegovina: A Tradition Betrayed.* New York: Columbia University Press, 1995.

Dragnich, Alex N. *Serbia's Historical Heritage.* Boulder, CO: East European Monographs, 1994.

Fine, John V. A., Jr. *The Early Medieval Balkans: A Critical Survey from the Sixth to the Late Twelfth Century.* Ann Arbor: University of Michigan Press, 1983.

———. *The Late Medieval Balkans: A Critical Survey from the Late Twelfth Century to the Ottoman Conquest.* Ann Arbor: University of Michigan Press, 1987.

Florescu, Radu, and Raymond T. McNally. *Dracula, Prince of Many Faces: His Life and His Times.* Boston: Little, Brown, 1989.

Georgescu, Vlad. *The Romanians: A History.* Translated by Alexandra Bley-Vroman. Rev. ed. Columbus: Ohio State University Press, 1991.

Glenny, Misha. *The Balkans: Nationalism, War and the Great Powers, 1804-1999.* London: Penguin, 2000.

———. *The Fall of Yugoslavia: The Third Balkan War.* New York: Penguin, 1992.

Goldstein, Ivo. *Croatia: A History.* Translated by Nikolina Jovanovic. Montreal: McGill-Queen's University Press, 1999.

Guldescu, Stanko. *The Croatian-Slavonian Kingdom, 1526-1792.* The Hague: Mouton, 1970.

———. *History of Medieval Croatia.* The Hague: Mouton, 1964.

Hall, Derek. *Albania and the Albanians.* London: Pinter, 1994.

Hitchins, Keith. *Rumania, 1866-1947.* Oxford: Clarendon Press, 1994.

———. *The Rumanians, 1774-1866.* Oxford: Clarendon Press, 1996.

Hupchick, Dennis P. *The Balkans: From Constantinople to Communism.* New York: Palgrave, 2002.

———. *The Bulgarians in the Seventeenth Century: Slavic Orthodox Society and Culture under Ottoman Rule.* Jefferson, NC: McFarland, 1993.

———. *Conflict and Chaos in Eastern Europe.* New York: St. Martin's Press, 1995.

———. *Culture and History in Eastern Europe.* New York: St. Martin's Press, 1994.

Hupchick, Dennis P., and Harold E. Cox. *The Palgrave Concise Historical Atlas of Eastern Europe.* Rev. and updated ed. New York: Palgrave, 2001.

Inalcik, Halil. *The Ottoman Empire: The Classical Age, 1300-1600.* Translated by Norman Itzkowitz and Colin Imber. London: Weidenfeld & Nicolson, 1973.

Jacques, Edwin E. *The Albanians: An Ethnic History from Prehistoric Times to the Present.* Jefferson, NC: McFarland, 1995.

Jelavich, Barbara. *History of the Balkans.* 2 vols. Cambridge: Cambridge University Press, 1985.

Jelavich, Charles, and Barbara Jelavich. *The Establishment of the Balkan National States, 1804-1920*. Seattle: University of Washington Press, 1977.

Judah, Tim. *The Serbs: History, Myth and the Destruction of Yugoslavia*. New Haven, CT: Yale University Press, 1997.

Kolarz, Walter. *Myths and Realities in Eastern Europe*. London: Lindsay Drummond, 1946.

Lampe, John R. *Yugoslavia as History: Twice There Was a Country*. Cambridge: Cambridge University Press, 1996.

Lang, David M. *The Bulgarians: From Pagan Times to the Ottoman Conquest*. Boulder, CO: Westview Press, 1976.

MacDermott, Mercia. *A History of Bulgaria, 1393-1885*. New York: Praeger, 1962.

Macfie, A. L. *The End of the Ottoman Empire, 1908-1923*. London: Longman, 1998.

Magocsi, Paul Robert, and Geoffrey J. Matthews. *Historical Atlas of East Central Europe*. Seattle: University of Washington Press, 1993.

Malcolm, Noel. *Bosnia: A Short History*. New York: New York University Press, 1995.

———. *Kosovo: A Short History*. New York: New York University Press, 1998.

Matthew, Donald. *Atlas of Medieval Europe*. Oxford: Oxford University Press, 1983.

McCarthy, Justin. *The Ottoman Turks: An Introductory History to 1923*. London: Longman, 1997.

Norwich, John J. *Byzantium*. 3 vols. New York: Knopf, 1992-96.

O'Ballance, Edgar. *Civil War in Bosnia, 1992-94*. New York: St. Martin's Press, 1995.

Obolensky, Dimitri. *The Byzantine Commonwealth: Eastern Europe, 500-1453*. New York: Praeger, 1971.

Osborne, R. H. *East-Central Europe: An Introductory Geography*. New York: Praeger, 1967.

Ostrogorsky, George. *History of the Byzantine State*. Translated by Joan M. Hussey. New Brunswick, NJ: Rutgers University Press, 1957.

Palmer, Alan. *The Decline and Fall of the Ottoman Empire*. New York: M. Evans, 1992.

Palmer, R. R., ed. *Atlas of World History*. New York: Rand McNally, 1957.

Pavlowitch, Stevan K. *A History of the Balkans, 1804-1945*. London: Longman, 1999.

Petrovich, Michael B. *A History of Modern Serbia, 1804-1918*. 2 vols. New York: Harcourt Brace Jovanovich, 1976.

Pinson, Mark, ed. *The Muslims of Bosnia-Herzegovina: Their Historic Development from the Middle Ages to the Dissolution of Yugoslavia*. 2nd ed. Cambridge, MA: Harvard Middle Eastern Monographs, 1996.

Poulton, Hugh. *Who Are the Macedonians?* 2nd ed. Bloomington: Indiana University Press, 2000.

Pounds, Norman J. G., and Robert C. Kingsbury. *An Atlas of European Affairs*. New York: Praeger, 1964.

Pribichevich, Stoyan. *Macedonia: Its People and History*. University Park: Pennsylvania State University Press, 1982.

Quataert, Donald. *The Ottoman Empire, 1700-1922*. Cambridge: Cambridge University Press, 2000.

Ramet, Sabrina P. *Balkan Babel: The Disintegration of Yugoslavia from the Death of Tito to the War for Kosovo*. 3rd ed. Boulder, CO: Westview Press, 1999.

———. *Nationalism and Federalism in Yugoslavia, 1962-1991*. 2nd ed. Bloomington: Indiana University Press, 1992.

Rogel, Carole. *The Slovenes and Yugoslavism, 1890-1918*. Boulder, CO: East European Monographs, 1977.

Rothenberg, Gunther E. *The Austrian Military Border in Croatia, 1522-1747*. Urbana: University of Illinois Press, 1960.

———. *The Military Border in Croatia, 1740-1881: A Study of an Imperial Institution*. Chicago, IL: University of Chicago Press, 1966.

Rothschild, Joseph. *East Central Europe between the Two World Wars*. Seattle: University of Washington Press, 1974.

Rothschild, Joseph, and Nancy M. Wingfield. *Return to Diversity: A Political History of East Central Europe Since World War II*. 3rd ed. New York: Oxford University Press, 2000.

Runciman, Steven. *A History of the First Bulgarian Empire*. London: G. Bell, 1930.

Sedlar, Jean W. *East Central Europe in the Middle Ages, 1000-1500*. Seattle: University of Washington Press, 1993.

Seton-Watson, R. W. *A History of the Roumanians*. New York: Shoe String Press, 1934.

Shepherd, William R. *Shepherd's Historical Atlas*. 9th ed. New York: Barnes & Noble, 1976.

Singleton, Fred B. *A Short History of the Yugoslav Peoples*. New York: Cambridge University Press, 1985.

Stavrianos, L. S. *The Balkans Since 1453*. New York: Holt, Rinehart & Winston, 1958.

Stoianovich, Traian. *Balkan Worlds: The First and Last Europe*. Armonk, NY: M.E. Sharpe, 1994.

Stokes, Gale. *The Walls Came Tumbling Down: The Collapse of Communism in Eastern Europe*. New York: Oxford University Press, 1993.

Sugar, Peter F. *Southeastern Europe under Ottoman Rule, 1354-1804*. Seattle: University of Washington Press, 1977.

Sugar, Peter F., and Ivo J. Lederer, eds. *Nationalism in Eastern Europe*. 2nd ed. Seattle: University of Washington Press, 1994.

Swain, Geoffrey, and Nigel Swain. *Eastern Europe Since 1945*. New York: St. Martin's Press, 1993.

Tanner, Marcus. *Croatia: A Nation Forged in War*. New Haven, CT: Yale University Press, 1997.

Treptow, Kurt W., ed. *A History of Romania.* Iaşi: Center for Romanian Studies, 1996.

Tsvetkov, Plamen. *A History of the Balkans: A Regional Overview from a Bulgarian Perspective.* 2 vols. Lewiston, NY: Edwin Mellen, 1993.

Vakalopoulos, Apostolos E. *The Greek Nation, 1453-1669: The Cultural and Economic Background of Modern Greek Society.* Translated by Ian Moles and Phania Moles. New Brunswick, NJ: Rutgers University Press, 1976.

———. *History of Macedonia, 1354-1833.* Translated by Peter Megann. Thessaloniki: Institute for Balkan Studies, 1973.

———. *Origins of the Greek Nation: The Byzantine Period, 1204-1461.* Translated by Ian Moles. New Brunswick, NJ: Rutgers University Press, 1970.

Vickers, Miranda. *Albania: A Modern History.* London: I.B. Tauris, 1994.

Vidal-Naquet, Pierre, and Jacques Bertin. *The Harper Atlas of World History.* Translated by Chris Turner, *et al.* New York: Harper & Row, 1987.

Whittow, Mark. *The Making of Byzantium, 600-1025.* Berkeley: University of California Press, 1996.

Wilkinson, H. R. *Maps and Politics: A Review of the Ethnographic Cartography of Macedonia.* Liverpool: University Press of Liverpool, 1951.

Williams, Joseph E., ed. *World Atlas.* Englewood Cliffs, NJ: Prentice-Hall, 1960.

Wolff, Robert L. *The Balkans in Our Time.* Cambridge, MA: Harvard University Press, 1956.

Woodhouse, Christopher M. *Modern Greece: A Short History.* 4th ed. London: Faber, 1986.

World History Atlas. Rev. ed. Maplewood, NJ: Hammond, 1993.

Index

This concise atlas is indexed according to map number. Non-italicized numbers refer to text pages; italicized numbers refer to map pages.

Abdülaziz, Ottoman sultan (1861-76), 26
Abdülhamid II, Ottoman sultan (1876-1909), 26, 27, 31
Abdülmecid I, Ottoman sultan (1839-61), 26
Achaia, 13, 14; *6, 13, 14, 17, 18*
Acheloös River, 3; *1, 40*
Adrianople, 13, 19; *6, 7, 8, 9, 11, 12, 13, 14, 18, 19. Also see* Edirne
Adriatic Sea, 1, 3, 5, 9, 13, 16, 17, 19, 34, 41, 43; *1, 41*
Aegean Sea, 1, 3, 8, 9, 13, 14, 28, 35, 43, 44; *1, 39, 40*
Aehrenthal, Alois, 31
Alba Iulia, *35, 36, 37, 38*
Albania, 2, 3, 5, 9, 14, 17, 18, 19, 24, 26, 33, 35, 40, 41, 42, 43, 44, 45, 46, 47, 48, 50; *2, 3, 4, 5, 8, 14, 17, 19, 21, 22, 24, 25, 26, 27, 28, 29, 30, 31, 32, 33, 34, 35, 36, 40, 41, 42, 43, 44, 45, 46, 47, 48, 49, 50*
Albanian Alps, 1, 3, 34; *1*
Albanians, 2, 4, 5, 17, 19, 29, 31, 33, 41, 43, 48, 49, 50; *4;* northern Ghegs, 4, 41; southern Tosks, 4, 41; Kosovar, 41, 50
Aleksandŭr, Ivan, Bulgarian *tsar* (1331-71), 18
Alexander, Greek king (1917-20), 40
Alexander I, Russian *tsar* (1801-25), 25
Alexander II, Russian *tsar* (1855-81), 27
Alexandroupolis, *40*
Alexandru the Good, Moldavian *voievod* (1400-32), 15
Alexandru, Nicolae, Wallachian *voievod* (1352-64), 15
Alia, Ramiz, Albanian Communist leader (1985-92), 48
Aliakmon River, 1, 3; *1, 40*
Allies (Anti-Axis), 44, 45
Americans, 27, 35, 36
Anatolia, 7, 9, 11, 12, 13, 14, 18, 19, 35, 40; *1*
Angelos, Alexios IV, Byzantine emperor (1203-4), 13
Angelos, Isaac II, Byzantine emperor (1185-95), 11, 13
Angelos, Theodore Doukas, Epirote prince (1214-30), 13
Anjou, Charles I de, Sicilian king (1262-85), 14
Anjou, Charles Robert I de, Hungarian king (1301-42), 15
Anjou, Louis I de, Hungarian and Polish king (1342-82), 15, 16
Antonescu, Ion, 42, 44
Aquileia, *6*
Arad, *36, 37, 38*
Archipelago, duchy of the, 14
Argeş River, *15, 37*
Arkadioupolis, *8*
Armenians, 19, 20, 22, 31
Arta, *12, 13, 14, 18, 19, 40, 46*
Asen, Ivan, Bulgarian *tsar* (1187-96), 11
Asen, Ivan II, Bulgarian *tsar* (1218-41), 13, 14
Asen, Petŭr, Bulgarian *tsar* (1196-97), 11
Asia, *6*
Asparuh, Bulgar *han* (d. 701), 7, 8
Athens, 2, 14, 25, 42, 43, 44, 46; *2, 6, 7, 8, 9, 11, 12, 13, 14, 17, 18, 19, 21, 22, 24, 25, 26, 27, 28, 29, 31, 33, 34, 35, 40, 42, 43, 44, 45, 46, 47, 48*
Attica, *1*
Austria, 31, 35, 37, 43; *2, 3, 4, 5, 35, 36, 42, 43, 44, 45, 47, 48. Also see* Austria-Hungary
Austria-Hungary, 26, 28, 29, 30, 31, 32, 34, 37; *26, 27, 28, 29, 31, 32, 33, 34. Also see* Austria, Habsburg (Austrian) Empire, *and* Hungary
Avars, 4, 7, 10; *6, 7*
Axis Powers, 42, 43, 44
Ayans, 24
Bajram Curri, *41, 50*
Baldwin I, Latin emperor of Constantinople (1204-5), 13
Balik, 18
Balkan Campaign (1941), 43, 44
Balkan Conferences (1931-33), 40, 42
Balkan Entente (1934), 39, 40
Balkan League (1912), 33
Balkan Mountains, 1, 3, 8, 11, 19, 27, 28; *1*
Balkan Wars (1912-13), 30, 33, 34, 35, 41; *33*
"Balkanization," 28
Banat, 5, 24, 35; *1, 15, 23, 24, 25, 26, 27, 28, 29, 31, 32, 34, 35, 37, 38, 44, 48*
Banja Luka, *23, 32, 36, 49*
Banovinas, 36
Bansko, *30*
Bar, *10, 41*
Bari, *9*
Basarab, Wallachian *voievod* (*ca.* 1310-52), 15
Başıbazuks, 27
Basil II *Boulgaroktonos,* Byzantine emperor (963-1025), 9, 10, 11
Battenberg, Aleksandŭr (Alexander) I, Bulgarian prince (1879-86), 29
Bavaria, *25*
Bayezid I the Thunderbolt, Ottoman sultan (1389-1402), 19
Béla I, Hungarian king (1061-63), 10
Belgrade (Singidunum), 2, 3, 6, 17, 21, 24, 25, 31, 32, 34, 35, 36, 43, 45; *2, 6, 7, 8, 9, 10, 11, 12, 13, 14, 15, 16, 17, 18, 19, 21, 22, 23, 24, 25, 26, 27, 28, 29, 31, 32, 33, 34, 35, 36, 37, 38, 42, 43, 44, 45, 47, 48, 49*
Belgrade, pronouncement of (1918), 35, 36
Beli Drin River, *41, 50*
Benevento, *8, 9*
Berat, *41*
Berlin, and congress of (1878), 28, 29, 30, 31, 35, 44
Bessarabia, 4, 27, 28, 29, 34, 35, 37, 42, 44, 45, 47; *1, 14, 15, 18, 19, 21, 22, 25, 26, 27, 28, 29, 31, 34, 35, 37, 42, 44, 45, 47, 48. Also see* Moldova
Bihać, *23, 32, 43, 48, 49*
Biograd, 10; *8, 9, 10*
Bismarck, Otto von, 28

Bistrița, *38*
Bitola, 33, 34; *24, 29, 30, 33, 34, 41*
Blače, *50*
"Black Hand" ("Union or Death"), 32
Black Sea, 1, 8, 9, 14, 24, 25; *1, 15, 37, 39, 40*
Blaj, *38*
Bœtia, *24*
Bogdan I, Moldavian *voievod* (1359-ca. 1365), 15
Bogomilism/Bogomils, 9, 11, 16, 17
Bohemia, *13*
Bolsheviks, 35, 39
Boris I, Bulgar prince (852-89), 8
Boris II, Bulgarian *tsar* (967-71), 9
Boris III, Bulgarian king (1918-43), 35, 39, 42, 44
Bosilegrad, *39*
Bosna River, 3, 16; *1, 10, 16, 23, 32*
Bosnia, 3, 5, 10, 11, 13, 16, 17, 19, 23, 26, 28, 32, 34, 35, 44, 49, 50; *9, 10, 11, 12, 13, 14, 16, 17, 18, 19, 21, 22, 23, 24, 25, 26, 27, 32, 49*. *Also see* Bosnia-Hercegovina
Bosnia-Hercegovina, 2, 3, 5, 26, 27, 28, 29, 30, 31, 32, 43, 45, 49, 50; *2, 3, 4, 5, 28, 29, 31, 32, 33, 34, 35, 44, 45, 47, 48, 49*. *Also see* Bosnia *and* Hercegovina
Bosnia-Hercegovina, Serbian Republic of, 49
Bosnian church, 16
Bosnians, 2, 4, 16, 19; *4*
Bosphorus Strait, 6, 7, 12, 14, 20, 26; *1, 20*. *Also see* Straits
Bouillon, Baldwin de, 12
Bouillon, Godfrey de, 12
Brăila, *15, 34, 37*
Braničevo (Viminacium), *6, 8, 14, 17, 18*
Branimir, Croatian ruler (879-92), 10
Brașov, *15, 37, 38*
Bratislava, *18, 21, 35, 42, 43, 44, 45, 47, 48*
Brčko, *49*
Bribar, *10*
Britain, 24, 25, 26, 27, 29, 31, 35, 39, 40, 41, 43, 44, 45, 46, 47
British, 24, 34, 35, 40, 43, 44, 45, 46
Brod, *23*
Bucharest, 2, 25, 27, 29, 33, 34, 42, 47, 48; *2, 19, 21, 22, 24, 25, 26, 27, 28, 29, 31, 33, 34, 35, 37, 39, 42, 43, 44, 45, 47, 48*
Bucharest Treaty: of 1812, 25; of 1886, 29; of 1913, 33; of 1918, 34
Buchlau, meeting of (1908), 31
Budapest (Buda and Pest), 21; *2, 19, 21, 22, 24, 25, 26, 27, 28, 29, 31, 34, 35, 42, 43, 44, 45, 47, 48*
Budapest, convention of (1877), 27, 28
Bukovina, 15, 35, 42, 45, 47; *15, 24, 37, 38, 42*
Bulgaria, 2, 3, 5, 7, 8, 9, 10, 11, 13, 14, 15, 17, 18, 19, 22, 24, 26, 27, 29, 30, 31, 33, 34, 35, 39, 40, 42, 43, 44, 45, 48; *2, 3, 4, 5, 7, 8, 9, 11, 12, 13, 14, 15, 17, 18, 19, 21, 22, 24, 25, 26, 27, 28, 29, 30, 31, 33, 34, 35, 36, 37, 39, 40, 42, 43, 44, 45, 46, 47, 48*
Bulgarian Agrarian Union party, 39, 42, 45
Bulgarian ("April") Uprising (1876), 26, 27
"Bulgarian Church Question" (1860-72), 26, 27, 30
Bulgarian Exarchate, 26, 27, 30
Bulgarians/Bulgars, 2, 4, 5, 7, 8, 9, 12, 13, 14, 17, 18, 19, 22, 26, 27, 28, 29, 30, 33, 34, 35, 37, 42, 44, 48; *4, 7*
Burgas, *39, 45, 48*
Bursa, 19; *18, 19, 40*
Butrint, *18*
Byzantine (East Roman) Empire (Byzantium), 1, 4, 5, 6, 7, 8, 9, 10, 11, 12, 13, 14, 15, 16, 17, 18, 19, 20, 30; *6, 7, 8, 9, 10, 11, 12, 14, 17, 18*
Capodistrias, John, 25
Carol (Karl) I, Romanian prince and king (1866-1914), 29, 37
Carol II, Romanian king (1930-40), 37, 42
Carpathian Mountains, 1, 3, 4, 15, 37; *1*
Catalan Grand Company, 14
Çatalca, 33; *33*
Catherine II the Great, Russian *tsarina* (1762-96), 24
Ceaușescu, Nicolae, Romanian Communist leader (1965-89), 47, 48
Central Powers (Alliance), 29, 31, 32, 34, 35, 36, 37
Cetatea Alba, *15, 19, 21*
Cetinje, *22, 23, 25, 26, 27, 28, 29, 31, 32, 33, 34, 35, 36*
Četniks, 44, 45
Charlemagne, Frank king and Holy Roman emperor (771-814), 10
Charles I, Habsburg emperor (1916-22), 34
Charles V, Holy Roman emperor (1519-56), 21
Chernivtsi, *37*
China (Maoist), 47
Christianity, 5, 6, 9, 11, 12, 25, 27, 28, 29, 31, 32; Armenian (Gregorian Monophysite), 22; Orthodox, 2, 4, 5, 6, 8, 11, 12, 13, 14, 15, 16, 17, 20, 22, 23, 24, 25, 26, 27, 28, 31, 32, 36, 37, 41, 44; *22*; Protestant, 2, 5; Roman Catholic, 2, 5, 10, 11, 12, 13, 14, 15, 16, 17, 22, 31, 32, 36, 37, 41, 44; *22*
Circassians, 27
Civilization: Eastern European (Orthodox), 5, 13, 15, 17, 20, 23, 32, 40, 46; *5*; Islamic (Muslim), 5, 20, 22, 24, 26, 32, 41, 48; *5*; Western European (Roman Catholic/Protestant), 5, 13, 23, 24, 32, 46; *5*
Cluj, *15, 19, 21, 22, 27, 28, 29, 31, 34, 35, 37, 38, 42, 43, 44, 45, 47, 48*
Codreanu, Corneliu, 37
Committee for Union and Progress (CUP), 31
Communism (Communists), 38, 39, 42, 44, 45, 46, 47, 48, 49, 50; *45, 47, 48*. *Also see* Bolsheviks
Communist Information Bureau (Cominform), 45, 47; *45*. *Also see* Communism
Communist International (Comintern), 45. *Also see* Communism

Constanţa (Tomi), *6, 27, 28, 29, 31, 33, 34, 37, 39, 45, 47, 48*
Constantine, Greek king (1913-17; 1920-22), 34, 40
Constantine I the Great, Roman emperor (306-37), 6
Constantinople (Byzantion), 6, 7, 8, 10, 11, 12, 13, 14, 17, 18, 19, 20, 22; *6, 7, 8, 9, 11, 12, 13, 14, 17, 18, 19, 20*. *Also see* Istanbul
Constantinople, (Greek) patriarchate of, 8, 11, 15, 17, 18, 20, 22, 25, 26, 30; *6, 22*
Corfu, 34; *13, 18, 19, 21, 26, 27, 28, 29, 33, 34, 40, 41, 42, 43, 44, 45, 46, 48*
Corfu, declaration of (1917), 34, 35, 36
Corinth, 3, 17; *6, 7, 9, 40, 43*
Çorovodë, *41*
Council of Mutual Economic Assistance (Comecon) (1949), 47
Craiova, and treaty of (1940), 42; *35, 37, 39, 42, 43, 45*
Crete, 13, 29, 43, 44; *1, 9, 11, 12, 13, 14, 17, 18, 19, 21, 22, 24, 25, 26, 27, 28, 29, 31, 34, 35, 40, 42, 43, 44, 45, 46, 47, 48*
Crimea, and Crimean War (1853-56), 24, 25, 26, 37
Criş River, *38*
Crişana, *35, 37, 38*
Črnojević, Arsenije III, Peć patriarch, 23
Croat Peasant party, 36
Croatia, 2, 3, 5, 8, 10, 13, 16, 21, 23, 32, 35, 36, 43, 44, 45, 49, 50; *2, 3, 4, 5, 8, 9, 10, 11, 12, 13, 14, 16, 17, 18, 19, 21, 22, 23, 24, 25, 26, 27, 28, 29, 31, 32, 34, 35, 44, 45, 47, 48, 49*
Croatian-Slavonian Military Border *(Krajina)*, 23, 25; *21, 22, 23*
Croats, 2, 4, 5, 7, 10, 11, 12, 23, 25, 32, 34, 35, 36, 42, 43, 44, 45, 49; *4, 7*
Crusaders, 11, 12, 13, 19
Crusades, 5, 11, 12, 13, 16, 19; *12*
Cumans, 4, 11, 12, 15, 17; *11, 12*
Curtea de Argeş, 15; *14, 15*
Cuza, Alexander, Romanian prince (1859-66), 37
Cyclades Islands, *25, 27, 28, 33, 40*
Cyprus, 28, 29
Cyril and Methodios, 8
Czechoslovakia, 41, 42; *35, 37, 45, 47, 48*
Czechs, 4, 23, 32
Dacians (Daks), 4, 38
Dakovica, *50*
Dalmatia, 3, 5, 10, 16, 21, 25, 26, 35, 36, 41, 43; *1, 6, 16, 19, 21, 22, 23, 24, 25, 26, 27, 28, 29, 31, 32, 33, 34, 35, 44, 49;* Croatian, 10; *10*
Dandolo, Enrico, Venetian *doge* (1193-1205), 13
Danube River, 1, 3, 4, 6, 7, 8, 9, 11, 15, 17, 18, 19, 21, 23, 24, 29, 37; *1, 2, 3, 4, 5, 6, 7, 8, 9, 10, 11, 12, 13, 14, 15, 16, 17, 18, 19, 21, 22, 23, 24, 25, 26, 27, 28, 29, 31, 32, 33, 34, 35, 36, 37, 38, 39, 42, 43, 44, 45, 47, 48, 49*
Danubian Basin, 8, 21
Dardanelles Strait, 26; *1*. *Also see* Straits
Dayton (Ohio) Peace Accord (1995), 49
Debar, *30, 41*
Debrecen, *15, 37, 38*
Dečani, *50*
Dečanski, Stefan Uroš III, Serbian king (1321-31), 17, 18
Diagonal Highway *(Via Militaris)*, 6, 12; *6*
Didymoteichon, *24*
Dimitrijević, Dragutin ("Apis"), 32
Dimitrov, Georgi, Bulgarian Communist leader (1946-49), 45
Dinaric Alps, 1, 3, 16; *1*
Disraeli, Benjamin, 27
Djakovica, *41*
Djakovo, *16*
Djed (grandfather), 16
Djurdjevac, *23*
Dniester River, *1, 15, 37*
Dobrotitsa, Bulgarian "Dobrudzhan" despot (ca. 1366-85), 18
Dobrudzha, 9, 27, 28, 33, 34, 35, 42, 45; *1, 8, 14, 15, 18, 19, 21, 22, 24, 25, 26, 27, 28, 29, 31, 33, 34, 35, 37, 39, 42, 44*
Dodecanese Islands, *33, 34, 35, 40, 42, 43, 44*
Doukas, Alexios V, Byzantine emperor (1204), 13
Drava River, 1, 3, 10; *1, 10, 16, 23, 32, 36, 49*
Drin River, 1, 3; *1, 10, 41, 50*
Drin i Zi River, *41, 50*
Drina River, 3; *1, 10, 16, 23, 32, 36*
Držislav, Stjepan, Croatian king (969-97), 10
Dubrovnik, 10, 11, 13, 16, 17; *7, 8, 9, 10, 11, 12, 13, 14, 16, 17, 18, 19, 21, 22, 23, 24, 25, 26, 27, 28, 29, 31, 32, 33, 34, 35, 36, 41, 42, 43, 44, 45, 47, 48, 49*
Duklja, *9, 10, 11, 14, 16, 17*
Durrës (Dyrrachion), 6, 9, 12, 13, 41; *6, 7, 8, 9, 11, 12, 13, 14, 17, 18, 19, 21, 22, 24, 26, 27, 28, 29, 31, 33, 34, 35, 40, 41, 42, 43, 44, 45, 48*
Dušan, Stefan Uroš IV, Serbian *car* (1331-55), 16, 17, 18, 19, 30
Eastern Rumelia, 28, 29; *28, 29*
"Eastern Question," 26
Edirne, 19, 27, 33, 35; *21, 22, 25, 26, 27, 28, 29, 31, 33, 34, 35, 39, 40, 42, 43, 44, 45, 46, 47, 48*. *Also see* Adrianople
Edirne (Adrianople), treaty of (1827), 25
Elbasan, *41*
Enez, *33*
Entente Powers (Alliance), 29, 31, 32, 34, 35, 39, 40
Enver Paşa, 31
Epidavros, *25*
Epiros, 9, 13, 14, 17, 18, 24, 29, 33, 35, 43, 44; *1, 6, 8, 13, 14, 17, 18, 19, 21, 22, 24, 25, 26, 27, 28, 29, 31, 33, 35, 41, 44, 46*

Esztergom, *9, 11, 12, 13, 14*
"Ethnic cleansing," 49
Eubœa, *14, 17, 18, 19, 24*
Eugene of Savoy, 24
European Community/Union (EC/U), 46, 49, 50
External Macedonian Revolutionary Organization (EMRO), 30
Fascists, 37, 40, 41, 42
Ferdinand, Romanian king (1914-27), 37
Ferdinand I, Bulgarian prince and king (1887-1918), 31, 33, 34
Ferdinand I, Hungarian king and Holy Roman emperor (1526-64), 21, 23
Fier, *41*
Filov, Bogdan, 44
Flor, Roger de, 14
Florina, *30, 34, 40, 41, 46*
"Fourteen Points" (1917), 35
France, 12, 21, 24, 25, 29, 31, 34, 35, 37, 39, 40, 41
Francis Ferdinand, 32, 37
Franciscans, 16
Franks, kingdom of, *7, 8. Also see* Holy Roman Empire
Frederick I Barbarossa, Holy Roman emperor (1152-90), 12
French, 12, 24, 25, 34, 35, 40, 41
Galata, *20*
Galaţi, *25, 26*
Gallipoli, 18, 19, 34; *12, 19, 34*
George II, Greek king (1922-23; 1935-47), 40, 43, 44, 46
Georgiev, Kimon, Bulgarian *zveno*-Communist leader (1944-46), 45
Gepids, *6*
Germans, 2, 4, 8, 12, 23, 36, 37, 38, 39, 42, 43, 44; *4;* Saxon, 16, 38
Germany, 26, 28, 29, 31, 34, 36, 38, 39, 40, 41, 42, 43, 44; *26, 27, 28, 29, 31, 42, 43, 44, 48. Also see* West Germany
Gheorghiu-Dej, Gheorghe, Romanian Communist leader (1953-65), 47
Giurgiu, *15, 37, 39*
Gjirokastër, *41*
Glasnost, 48
Glina, *49*
Gnjilane, *50*
Golden Horde, 14; *14, 15, 17, 18*
Golden Horn, 20; *20*
Goražde, *49*
Gorbachev, Mihail, Soviet leader (1985-92), 48, 49
Gorna Dzhumaya (Blagoevgrad), *30, 39*
Gostivar, *50*
Goths, 4, 6
Gračanica, *50*
Graz, *36*
Great Depression, 42
"Great Idea" *(Megale idaia)*, 30
Great Maina, 14
Great Moravia, 8; *8*
Great Schism (1054), 11, 12, 13, 17
Greece, 1, 2, 3, 5, 8, 18, 19, 24, 25, 26, 27, 28, 29, 30, 33, 34, 35, 39, 40, 42, 43, 44, 45, 46; *2, 3, 4, 5, 25, 26, 27, 28, 29, 30, 31, 33, 34, 35, 39, 40, 41, 42, 43, 44, 45, 46, 47, 48*
Greek Civil War (1947-49), 46, 47; *46*
Greek (Communist) Democratic Army, 46
Greek National Liberation Front/National Popular Party of Liberation (EAM/ELAS), 46
Greeks, 2, 4, 5, 17, 19, 20, 22, 25, 26, 27, 28, 29, 30, 33, 35, 37, 40, 43, 44, 46, 47; *4;* "Slavophone," 40, 46. *Also see* Phanariotes
Grevena, *30*
Groza, Petru, Romanian agrarian-Communist leader (1945-52), 45
Gypsies, 2, 4, 5
Habsburg (Austrain) Empire (the Habsburgs), 1, 21, 23, 24, 25, 26, 27, 28, 29, 31, 32, 34, 35, 36, 37, 38; *21, 22, 23, 24, 25. Also see* Austria-Hungary
Hagia Sophia (Aya Sofya), 6, 20
Heraclea, *6*
Herakleios, East Roman/Byzantine emperor (610-41), 7
Herceg-Bosna, Croat republic of, 49; *49*
Hercegovina, 3, 16, 26, 28, 35, 44, 49; *19, 23, 25, 26, 27, 32. Also see* Hum
Hesychism, 18
Hilandar Monastery, 11, 17
Hitler, Adolf, German Nazi leader (1933-45), 36, 38, 42, 43, 44
Holy Roman Empire, 11, 21; *9, 10, 11, 12, 13, 14, 16, 17, 18, 19, 24. Also see* Franks, kingdom of
Hotin, *15, 37*
Hoxha, Enver, Albanian Communist leader (1944-85), 45, 47
Hum, 16; *10, 14, 16, 17*
Hunedoara, *15, 38*
Hungarians (Magyars), 2, 4, 8, 10, 15, 16, 17, 18, 21, 23, 32, 36, 37, 38, 43, 44, 47, 48; *4, 8*
Hungary, 10, 11, 13, 14, 15, 16, 17, 19, 21, 23, 24, 35, 36, 37, 38, 42, 43, 47, 48; *2, 3, 4, 5, 9, 10, 11, 12, 13, 14, 15, 16, 17, 18, 19, 21, 22, 23, 24, 25, 35, 36, 37, 38, 42, 43, 44, 45, 47, 48, 49. Also see* Austria-Hungary, Habsburg (Austrian) Empire, *and* Royal Hungary
Hunyadi, János. 19
Iaşi, 34; *15, 22, 24, 25, 26, 27, 28, 29, 31, 34, 35, 37, 42, 43, 44, 45, 47, 48*
Ibar River, *50*
Ignatiev, Nicholas, 26
Ikaria, *46*
Illyria *(Illyricum)*, 4; *6*
"Illyrian Provinces," 25, 31; *25*

Innocent III (1198-1216), pope, 11, 12, 13
Internal Macedonian Revolutionary Organization (IMRO), 30, 39, 40, 42
Ioannina, *11, 17, 19, 21, 22, 24, 25, 26, 27, 28, 29, 31, 33, 34, 35, 40, 41, 42, 43, 44, 45, 46, 47, 48*
Ionian Islands, 24, 43; *40*
Ionian Sea, *40*
Iraklion, *40, 43, 46*
Iron Gates, 1; *1*
"Iron Guard" ("Legion of the Archangel Michael"), 37, 42
Iskŭr River, 1, 3; *1, 39*
Islam, 2, 11, 31, 32
Istanbul, 19, 20, 22, 24, 25, 26, 27, 33, 35, 42; *2, 21, 22, 24, 25, 26, 27, 28, 29, 31, 33, 34, 35, 40, 42, 43, 44, 45, 46, 47, 48*. *Also see* Constantinople
Istanbul, conference of (1876), 26
Istria, 25, 41; *1, 14, 18, 19, 21, 22, 24, 26, 27, 28, 29, 31, 34, 35, 44*
Italians, 4, 23, 35, 36, 40, 41, 42, 43, 44, 47
Italy, 6, 11, 14, 29, 33, 34, 40, 41, 42, 43, 44; *1, 2, 3, 4, 5, 6, 26, 27, 28, 29, 31, 33, 34, 35, 36, 41, 42, 43, 44, 45, 47, 48*
Izetbegović, Alija, Bosnian-Hercegovinian Muslim president (1991-present), 49
Izmaïl, *37*, 40; *34, 35, 40, 46*
Izvolsky, Alexander, 31
Jagiełło, Louis II, Hungarian king (1516-26), 21
Jajce, 45; *16, 23, 32, 49*
Janissaries, 19, 20, 21, 25, 26
Jews, 2, 4, 20, 22, 37, 42, 44
Julian Alps, 1, 3; *1*
Justinian I the Great, Roman emperor (527-65), 6, 20
Justinian II, Byzantine emperor (705-11), 8
Kačanik, *50*
Kálmán I, Hungarian king (1095-1114), 10
Kalocsa, 16; *9, 11, 22*
Kamenica, *50*
Kantakouzenos, John VI, Byzantine emperor (1347-54), 18
Karditsa, *46*
Kaloyan, Bulgarian king (1197-1207), 11, 13, 17
Kantakuzenos, John VI, Byzantine emperor (1347-54), 19
Karadjordjević, Aleksandr I, Yugoslav king (1921-34), 35, 36
Karadjordjević, Pavel, Yugoslav prince-regent (1934-41), 36, 43
Karadjordjević, Petr I, Serbian king (1903-21), 31, 32
Karadjordjević, Petr II, Yugoslav king (1934-41), 36, 43
Karadžić, Radovan, 49
Karlobag, *23*
Karlovac, *23*
Karlovo, *39*
Karpenision, *46*
Kastoria, *27, 28, 30, 40, 41*
Kastriotis (Skanderbeg), George, 19
Kavajë, *41*
Kavala, 33, 43; *27, 28, 30, 33, 34, 35, 39, 40, 42, 44, 45, 46, 48*
Kavarna, *18*
Kemal, Mustafa (Atatürk), 31, 35, 40
Khalkis, *40*
Khrushchev, Nikita, Soviet leader (1953-64), 47
Kievan Russia, 9; *9, 11, 12, 13, 14, 17, 18*. *Also see* Russia
Kilia, *15*
Kingdom of Serbs, Croats, and Slovenes, 35, 39, 41; *35*. *Also see* Yugoslavia
Kishinev, *35, 37, 42, 43, 44, 45, 48*
Klagenfurt, *35*
Knin, 49; *48, 49*
Komnenos, Alexios I, Byzantine emperor (1081-1118), 11, 12
Komnenos, Manuel I, Byzantine emperor (1143-80), 16
Komnenos, Michael Angelos, Epirote prince (1204-14), 13
Konitsa, *30, 46*
Korçe, *30, 41*
Kosovë Liberation Army (KLA), 50
Kosovo, 5, 13, 23, 33, 35, 41, 43, 45, 49, 50; *1, 14, 17, 18, 19, 21, 22, 23, 24, 25, 26, 27, 28, 29, 30, 31, 32, 33, 34, 35, 41, 44, 45, 47, 48, 49, 50*
Kosovo Polje, and battle of (1389), 19, 32, 50; *19, 50*
Kotor, 16; *10, 17, 23, 32, 41*
Kotromanić, Stjepan, Bosnian *ban* (ca. 1318-53), 16
Krajina, *49*
Kralj Marko, 19
Kratovo, *30*
Kresimir III, Croatian king (1000-30), 10
Kriva Palanka, *39*
Križevci, *23*
Krk Island, *10*
Krujë, *41*
Krum, Bulgar *han* (808-14), 8
Kuber, Bulgar chieftan (ca. 675-ca. 88), 7; *7*
Kubrat, Bulgar *han* (605-65), 7
Kukës, *41, 50*
Kulin, Bosnian *ban* (ca. 1180-1204), 11, 16
Kumanovo. *50*
Kupa River, *10*
Kuršumilja, *50*
Kutrigurs, *6*
Kyuchuk Kainardzha, and treaty of (1774), 24; *24*
Kyustendil (Velbuzhd), 17, 18; *14, 17, 18, 30, 39*
Lamia, *40, 46*
Larissa, *6, 8, 9, 13, 14, 17, 18, 19, 21, 22, 24, 26, 27, 28, 29, 31, 33, 34, 40, 42, 45, 46, 48*
Laskaris, Theodore I, Nicæan emperor (1204-22), 13
László I, Hungarian king (1077-95), 10
Latin Empire, 13, 14; *13*
Latins, 14
Lausanne, treaty of (1923), 35, 40

Lazar, Serbian prince (1371-89), 19
League of Nations, 38, 39
Lech, *19*
Lekapena, Maria. 9
Lenin, Vladimir, Soviet leader (1917-23), 39
Leopold I, Holy Roman emperor (1658-1705), 23
Leposavić, *50*
Lesbos, *46*
Leskovac, *50*
Lezhë, *41*
Little Entente, 38
Ljubljana, 2; *2, 23, 25, 26, 27, 28, 29, 31, 32, 34, 35, 36, 42, 43, 44, 45, 47, 48, 49*
Lombards, *6, 7*
London: conference of (1830), 25; of (1913), 33, 41
Lovech, *39*
Lüleburgaz, 33; *33, 39*
Lycus River, *20*
Lyon, Catholic church council of (1274), 14
Macedonia, 2, 3, 4, 5, 6, 7, 8, 9, 14, 17, 18, 19, 24, 26, 27, 28, 29, 30, 31, 33, 34, 35, 39, 40, 41, 42, 43, 44, 45, 46, 49, 50; *1, 2, 3, 4, 5, 6, 7, 8, 9, 11, 12, 13, 14, 17, 18, 19, 21, 22, 24, 25, 26, 27, 28, 29, 30, 31, 33, 34, 35, 39, 40, 41, 44, 45, 46, 47, 48, 50*
"Macedonian Question," 30, 39; *30*
Macedonians, 2, 4, 5, 30, 35, 36, 39, 42, 44, 46, 49; *4*
Maček, Vladko, 36
Maglaj, *49*
Mahmud II, Ottoman sultan (1808-39), 26
Maoism, 47
Maramureş, 15; *15, 35, 37, 38*
Maritsa River, 1, 3; *1, 39, 40*
Marmara Sea, *20, 39, 40*
Marshall Plan (1948), 47
Maurice, East Roman emperor (582-602), 7
Maximilian II, Holy Roman emperor (1564-76), 21
Mediterranean Sea, 1, 3, 13, 19, 24, 25, 26, 27, 29, 31, 44; *40*
Mehmed I the Restorer, Ottoman sultan (1413-21), 19
Mehmed II the Conqueror, Ottoman sultan (1451-81), 19, 20, 22
Mehmed VI, Ottoman sultan (1918-22), 35
Mesolongion, 25; *25*
Mesta/Nestos River, 1, 3; *1, 30, 39, 40*
Metaxas, John, Greek dictator (1936-41), 40, 43, 44
Methoni, *21*
Metsovon, *46*
Midhat, Ahmed Şefik, paşa, 26, 31
Mihai I, Romanian king (1940-47), 42, 44, 45
Mihailović, Dragoljub (Draža), 44, 45
Mihajlo, Zetan king (1051-81), 17
Miladinov brothers (Konstantin and Dimitŭr), 30
Millet, 20, 22, 23, 30, 49; Orthodox, 20, 22, 23, 26, 30; *22*
Milošević, Slobodan, Serbian and Yugoslav Communist leader (1987-2000), 48, 49, 50
Milutin, Stefan Uroš II, Serbian king (1282-1321), 17
Mitrovica, *50*
Mœsia, 7, 9; *1, 6, 7, 8, 9*
Mœsian Highway, 6; *6*
Mohács, and battle of (1526), 21, 23; *21, 23*
Moldavia, 3, 15, 20, 21, 24, 25, 27, 34, 37; *1, 14, 15, 17, 18, 19, 21, 24, 25, 35, 37, 38*
Moldova, 4; *2, 3, 4, 5*. *Also see* Bessarabia
Monemvasia, 14; *14, 17, 18, 19, 21*
Mongols (-Tatars), 4, 15, 18, 19
Montana, *39*
Montenegrins, 4, 5, 28, 35; *4*
Montenegro, 2, 3, 5, 9, 18, 26, 27, 28, 33, 34, 35, 36, 43, 45, 49, 50; *2, 4, 5, 19, 21, 22, 23, 24, 25, 26, 27, 28, 29, 31, 32, 33, 34, 35, 41, 44, 45, 47, 48, 49, 50*. *Also see* Zeta
Montferrat, Boniface di, 13
Morava River, 1, 3; *1, 36, 50*
Morava-Vardar Highway, 6; *6*
Moravians, *7*
Morea, 14, 24; *14, 17, 18, 19, 21, 22, 24*
Morosini, Pier, Latin (Catholic) patriarch of Constantinople, 13
Mostar, 49; *23, 26, 32, 48, 49*
Mount Athos, 11, 17; *17, 22, 30, 40*
Mrnjavčević, Vukašin, 18, 19
Murad I, Ottoman sultan (1360-89), 19
Murad II, Ottoman sultan (1421-51), 19
Mureş River, *15, 37, 38*
Muslims, 11, 12, 22, 24, 25, 26, 27, 31, 32, 40, 41, 44, 48, 49; Bosnian (Bosniaks), 2, 5, 35, 36, 44, 49
Mussolini, Benito, Italian fascist leader (1922-43), 40, 41, 42, 43
Mustafa III, Ottoman sultan (1757-74), 24
Mystras, 14; *13, 14, 17, 18, 19*
Naples (Neopolis), *6, 7, 9, 11, 12, 13, 14, 17, 18, 19, 21, 22, 24, 25, 26, 27, 28, 29, 31, 33, 34, 35, 42, 43, 44, 45, 47, 48*
Napoleonic Wars, 24, 25
Nauossa, *46*
Navarino, and battle of (1827), 25; *25*
Navpaktos, *19, 21*
Navplion, *25, 40*
Nazis (National Socialists), 36, 39, 41, 42, 44
Neamţ, *15*
Nedić, Milan, 43
Nemanja, Stefan I, Serbian prince (ca. 1167-96), 11, 17
Nemanja, Stefan II, Serbian prince (1196-1227), 17
Neopatras, duchy of, 14; *18*
Neretva River, 1, 3, 16; *1, 10, 16, 23, 32*
Nesebŭr (Mesembria), *6, 7, 8*
Neuilly, treaty of (1919), 35, 39
Nevrokop, *30, 39*

Nicæa, 13, 14, 17; *13*
Nicephoros I, Byzantine emperor (802-11), 8
Nikopol, 19; *15, 19, 39*
Nikopolis, *6*
Nin, *10*
Niš (Naissos), 17, 19, 34; *6, 7, 12, 13, 14, 17, 18, 19, 21, 22, 24, 25, 26, 27, 28, 29, 31, 33, 34, 35, 36, 42, 43, 45, 48, 50*
Niš, declaration of (1914), 34
Noli, Fan, 41
Norman Principalities, *12*
Normans, 10, 11, 12, 13
North Atlantic Treaty Organization (NATO), 46, 47, 50
Nova Gradiška, *23*
Novi Pazar, 33; *23, 26, 27, 28, 29, 31, 32, 50. Also see* Sandjak (of Novi Pazar)
Novi Sad, *23, 31, 32, 35, 36, 42, 43, 45, 47, 48, 49*
Novo Brdo, *50*
Obrenović, Aleksandr, Serbian king (1889-1903), 31, 32
Obrenović, Milan, Serbian prince and king (1868-89), 29
Obrenović, Miloš, Serbian prince (1830-39; 1858-60), 25
Odessa, *24, 37, 44*
Ogulin, *23*
Oğuzes, 4, 15; *9*
Ohrid (Lichnidos), 8, 9, 22, 30; *6, 7, 8, 9, 11, 12, 13, 14, 17, 18, 19, 21, 22, 24, 26, 27, 28, 29, 30, 31, 33, 35, 36, 40, 41, 42, 43, 44, 45, 48*
Ohrid, archbishopric-patriarchate of, 8, 9, 22, 30; *22*
Okučani, *49*
Olt River, *1, 15, 37, 38*
Oltenia, 24; *24*
Oradea, *37, 38*
Orhan I, Ottoman *emir* (1324-60), 19
Ormenion (Chernomen), battle of (1371), 19
Osijek, *32, 49*
Osman I, Ottoman *emir* (1281-1324), 19
Otočac, *23*
Otranto, *41*
Ottoman Empire (the Ottomans), 1, 5, 19, 20, 21, 22, 23, 24, 25, 26, 27, 28, 29, 30, 31, 33, 34, 35, 37, 38, 40, 41, 47; *18, 19, 20, 21, 22, 23, 24, 25, 26, 27, 28, 29, 31, 33, 34*
Pacta Conventa (1102), 10
Palaiologos, Andronicus II, Byzantine emperor (1282-1328), 14, 17, 18
Palaiologos, Andronicus III, Byzantine emperor (1328-41), 18
Palaiologos, Constantine XI, Byzantine emperor (1448-53), 20
Palaiologos, John V, Byzantine emperor (1341-76), 18, 19
Palaiologos, Michael VIII, Byzantine emperor (1259-82), 14
Pale, *49*
Pannonia, 3, 7, 8, 9, 10, 35, 38; *1, 6, 8;* Croatian, 10; *9, 10*
Pannonian Slavs, *7*
Papacy/pope(s), 10, 11, 12, 13, 14, 16, 17; *6*
Papagos, Alexander, 46
Papal States, 21; *11, 12, 13, 14, 17, 18, 19, 21, 22, 24, 25*
Papandreou, George, 46
Paris, treaty of (1947), 45
Paristrion, 9
Partisans (Communist), 44, 45, 46
Pašić, Nikola, 31, 36
Pasvanoğlu Osman Paşa, 24
Patras, *6, 7, 14, 17, 18, 19, 22, 25, 26, 27, 28, 29, 31, 33, 34, 40, 42, 43, 45, 46, 48*
Pavelić, Ante, 43, 44
Pechenegs, 4, 12, 15; *8, 9*
Peć, 17, 22; *17, 18, 22, 23, 24, 27, 28, 41, 50*
Peć, patriarchate of, 17, 22, 23; *22*
Pecs, *10, 16, 36*
Pedro III, Aragonese king (1276-85), 14
Peleponnese, 1, 3, 4, 7, 14, 25, 43; *1, 46*
Pella, *30*
Pera, *20*
Perestroika, 48
Pernik, *39*
Peshkopi, *41*
Peter the Hermit, 12
Petŭr Kresimir IV, Croatian king (1058-75), 10
Petrich, 9; *9*
Petrovaradin, *23, 24*
Petrović, Djordje (Karadjordje), 25
Petŭr I, Bulgarian *tsar* (927-67), 9
Phanar, 20; *20*
Phanariotes, 20, 25, 37, 47
Philike hetairia (Society of Friends), 25
Phokas, Nikephoros II, Byzantine emperor (963-69), 9
"Pig War" (1906-11), 31
Pindos Mountains, 1, 3; *1*
Pirgos, *40, 46*
Pirot, *39*
Plav, *50*
Pleven (Plevna), 27; *27*
Pliska, 8; *7, 8*
Ploieşti, 42, 44; *35, 37, 42, 43, 44, 45*
Plovdiv (Philippopolis), 28, 29; *6, 7, 8, 9, 11, 12, 14, 17, 18, 19, 21, 22, 25, 26, 27, 28, 29, 30, 31, 33, 34, 35, 39, 40, 42, 43, 44, 45, 46, 47, 48*
Podgorica, *2, 41, 42, 43, 44, 48, 49. Also see* Titograd
Podgradec, *41*
Podujevo, *50*
Poland, 15, 21, 37, 47; *9, 15, 17, 19, 21, 22, 24, 35, 37, 42, 44*
Poles, 15
Pomaks, 5
Pomorie, *7, 8*
Posveta (Enlightenment), 32
Potsdam, conference of (1945), 45

Požarevac, and treaty of (1718), 24; *24*
Prague, peace of (1562), 21
Preševo, *50*
Preslav, 8, 9; *8, 9*
Prespa, 9; *8, 9*
Preveza, *40*
Prilep, 18, 19; *18, 30, 41*
Princip, Gavril, 32
Priština, *27, 28, 29, 34, 41, 42, 45, 47, 48, 50*
Prizren, *30, 31, 34, 41, 48, 50*
Prussia, 23, 26
Prut River, 1, 3, 15, 45; *1, 15, 37*
Pukë, *41*
Radić, Stjepan, 36
Radimlja, *16*
Raška, 9, 11, 17; *10, 11, 12, 13, 14, 16, 17, 50*
Ratiaria, *6*
Ravenna, *7*
Red Army (Soviet), 44, 45
Reichstadt, agreement of (1876), 27, 28
Resen, *30*
Revolutions of 1848, 37, 38
Rhodes, *13, 21, 40, 45, 46, 47, 48*
Rhodope Mountains, 1, 3; *1*
Ribbentrop-Molotov Pact (1939), 42
Rijeka, *23, 25, 32, 34, 35, 36, 43, 44, 45, 47, 48, 49*
Rila Mountains, 1, 3; *1*
Rilski, Ivan, Bulgarian saint, 9
Robert College, 27
Romans, 4, 5, 6
Romania, 2, 3, 4, 5, 25, 27, 28, 29, 33, 34, 35, 37, 38, 39, 40, 42, 43, 44, 45, 47, 48; *2, 3, 4, 5, 25, 26, 27, 28, 29, 31, 33, 34, 35, 36, 37, 38, 39, 42, 43, 44, 45, 47, 48*
Romanian "Christmas Revolution" (1989), 48
Romanian National Salvation Front (NSF), 48
Romanian Principalities, 15, 20, 21, 25, 27, 37; *15. Also see* Moldavia *and* Wallachia
Romanians, 4, 5, 29, 33, 35, 37, 38, 42, 44, 48; *4*
Royal Hungary, 21, 23; *21, 22*
Rožaj, *50*
Rugova, Ibrahim, 50
Ruse, *37, 39*
Russia, 1, 23, 24, 25, 26, 27, 28, 29, 30, 31, 32, 33, 34, 36, 37, 49; *24, 25, 26, 27, 28, 29, 31, 34. Also see* Kievan Russia *and* Soviet Union (USSR)
Russians, 2, 4, 24, 26, 27, 28, 29, 31, 37, 38, 42
Russo-Turkish War: of 1806-7, 25; of 1809, 25; of 1828-29, 25; of 1877-78, 27, 29, 30, 37
Ruthenians, 4, 15
Salisbury, Robert of, 26
Samokov, *39*
Samos, *46*
Samuil, Bulgarian *tsar* (976-1014), 9, 10, 30
San Stefano, and treaty of (1878), 27, 28, 29, 30; *27*
Sandjak (of Novi Pazar), 28, 33; *28, 31, 32, 33*
Saracens, *9*
Sarajevo, 2, 32, 34, 49; *2, 16, 19, 21, 22, 23, 24, 25, 26, 27, 28, 29, 31, 32, 33, 34, 35, 36, 42, 43, 44, 45, 47, 48, 49*
Sarandë, *41*
Satu-Mare, *37, 38*
Sava (Rastko), Serbian saint, 11, 17
Sava River, 1, 3; *1, 10, 16, 23, 32, 36, 37, 49*
Selim I the Grim, Ottoman sultan (1513-20), 21, 22
Selim II the Sot, Ottoman sultan (1566-74), 21
Seljuk Empire (and principalities), 19; *13, 14, 17, 18, 19*
Senj, *23*
Senta, *23*
Serbia, 2, 3, 5, 11, 14, 15, 16, 17, 18, 19, 22, 23, 24, 25, 26, 27, 28, 29, 30, 31, 32, 33, 34, 35, 36, 43, 45, 48, 49, 50; *2, 4, 5, 9, 11, 12, 13, 14, 15, 16, 17, 18, 19, 21, 22, 23, 24, 25, 26, 27, 28, 29, 30, 31, 32, 33, 34, 35, 44, 45, 47, 48, 49, 50*
Serbian *"Krajina,"* 49
Serbo-Croat coalition, 32, 35
Serbs, 2, 4, 5, 7, 8, 9, 11, 12, 14, 16, 17, 18, 19, 22, 23, 25, 26, 28, 29, 30, 32, 33, 34, 35, 36, 43, 44, 45, 49, 50; *4, 7, 8, 10*
Şeriat (Islamic Sacred Law), 22
Serres, 18, 19; *11, 14, 17, 18, 30, 39, 40, 46*
Seven Slav Tribes, *7*
Severin, *15. Also see* Turnu-Severin
Sèvres, treaty of (1920), 35, 40
Shishman, Ivan, Bulgarian "Tŭrnovo" *tsar* (1371-93), 18, 19
Shishman, Mihail, Bulgarian *tsar* (1323-30), 18
Shkodër (Scodra), 33; *6, 10, 16, 18, 23, 27, 28, 29, 31, 33, 34, 35, 36, 40, 41, 42, 43, 45, 48*
Shkumbin River, 1, 3; *1, 41*
Shumen, *39, 48*
Šibenik, *23, 32*
Sibiu, *15, 24, 37, 38*
Sicilian Vespers (1282), 14
Sicily, 6, 13, 14; *1, 8, 11, 12, 13, 14, 17, 18, 19, 21, 22, 24, 25, 26, 27, 28, 29, 31, 33, 34, 35, 42, 43, 44, 45, 47, 48*
Sighişoara, *15*
Sigismund I, Hungarian king and Holy Roman emperor (1387-1437), 19
Silistra (Durostolon), *6, 7, 9, 15, 18, 24, 27, 28, 29, 33, 34, 35, 37, 39, 42, 43, 44, 45, 47, 48*
Simeon I, Bulgarian *tsar* (893-927), 8, 9, 10
Simonis, 17
Sinan, 20
Sirmium, 9
Sisak, *23*
Sitnica River, *50*
Skopje (Skupi), 2, 17, 30; *2, 6, 7, 13, 14, 17, 19, 21, 22, 25,*

26, 27, 28, 29, 30, 31, 33, 34, 35, 36, 40, 41, 42, 43, 44, 45, 46, 47, 48, 50
Slavonia, 3, 5, 7, 10, 23, 25, 35, 43, 49; *1, 10, 11, 13, 14, 16, 17, 18, 19, 21, 22, 23, 25, 26, 27, 28, 29, 31, 32, 34, 35, 49*
Slavs, 4, 7, 8, 9, 22, 30, 32, 34, 39, 40, 44, 46, 47; *6, 7;* Antes, 7; *6;* Slaveni, 7; *6*
Sliven, *39*
Slovakia, 23; *42, 43, 44*
Slovaks, 4, 23; *7*
Slovenes, 2, 4, 5, 25, 34, 35, 36, 45, 49; *4, 7, 8*
Slovenia, 2, 25, 35, 36, 43, 45, 49, 50; *2, 3, 4, 5, 23, 25, 26, 27, 28, 29, 31, 32, 34, 35, 44, 45, 47, 48, 49*
Slunj, *23*
Smederevo, *19*
Smolyan, *39*
Society of Liberty, 31
Sofia (Serdika), 2, 8, 9, 19, 29, 44, 45; *2, 6, 7, 8, 9, 11, 12, 13, 14, 17, 18, 19, 21, 22, 24, 25, 26, 27, 28, 29, 31, 33, 34, 35, 36, 39, 42, 43, 44, 45, 47, 48*
Sokollu, Mehmed, 22
Someş River, *37, 38*
Southern Bug River, *37*
Soviet Union (USSR), 38, 39, 41, 42, 43, 44, 45, 46, 47, 48; *35, 37, 42, 43, 44, 45, 47, 48. Also see* Russia
Sozopol (Sozopolis), *6*
Sparta, *40*
Split (Salona), *6, 7, 9, 10, 11, 12, 14, 16, 18, 19, 21, 22, 23, 26, 27, 28, 29, 31, 32, 36, 42, 43, 44, 45, 47, 49*
Sporazum, 36, 43; *36*
Srebrenica, *16, 23, 32, 49*
Srem (Sirmium), *6, 7, 12, 16*
Sremski Karlovci, and treaty of (1699), 23; *22, 23, 24*
Stalin, Joseph, Soviet Communist leader (1924-53), 45, 46, 47, 49
Stalingrad, battle of (1942-43), 44
Stamboliiski, Aleksandŭr, 35, 39, 42
Stambolov, Stefan, 29
Stara Zagora, *39*
Štip, *30*
Stefan Uroš V, Serbian *car* (1355-71), 18
Stjepan I, Croatian king (1030-58), 10
Ston, *16*
Straits, 26, 27, 28
Stratsimir, Ivan, Bulgarian "Vidin" *tsar* (1370-96), 18
Struga, *41*
Struma/Strymon River, 1, 3; *1, 30, 39, 40*
Strumica, *30, 33, 34, 39*
"Sublime Porte," 20
Suceava, 15; *14, 15, 18, 19, 21*
Süleyman I the Magnificent, Ottoman sultan (1520-66), 20, 21, 23
Šumadija, 3; *1*
Svetoslav, Teodor, Bulgarian *tsar* (1300-21), 18
Svishtov, *39*
Svyatoslav, Kievan ruler (962-72), 9
Szeged, *15, 36, 37, 38*
Székelys, 38; *4*
Szigetvar, 21; *21, 23*
Taigetos Mountains, 1, 3; *1*
Tanzimat, 26
Tepedenli Ali Paşa, 24
Tepelenë, *41*
Tervel, Bulgarian *han* (701-18), 8
Tetovo, *41, 50*
Thebes, *25, 40*
Thessaloniki, 6, 8, 9, 13, 17, 19, 27, 31, 33, 34, 35, 42, 43, 44; *6, 7, 8, 9, 11, 12, 13, 14, 17, 18, 19, 21, 22, 24, 25, 26, 27, 28, 29, 30, 31, 33, 34, 35, 39, 40, 42, 43, 44, 45, 46, 47, 48*
Thessaly, 3, 9, 14, 17, 18, 29; *1, 6, 8, 9, 13, 14, 17, 19, 21, 22, 24, 25, 26, 27, 28, 29, 46*
Thrace (Thracia), 3, 4, 7, 8, 9, 14, 18, 26, 27, 28, 33, 35, 39, 40, 43, 44; *1, 6, 8, 13, 14, 17, 18, 19, 21, 22, 24, 25, 26, 27, 28, 29, 31, 33, 34, 35, 39, 40, 44, 46, 48*
Timişoara, 48; *15, 35, 36, 37, 38, 42, 43, 44, 45, 48*
Tiranë, 2, 41; *2, 35, 36, 40, 41, 42, 43, 44, 45, 46, 47, 48*
Tîrgovişte, *15, 18, 19*
Tîrgu-Mureş, *37, 38*
Tisza River, 3; *1, 10, 15, 36, 37, 38*
Tito (Broz), Josip, Yugoslav Communist leader (1944-80), 44, 45, 46, 47, 48, 49, 50
Tito-Stalin split (1948), 46, 47, 49
Titograd, *45, 47. Also see* Podgorica
Tökés, László, 48
Tomislav, Croatian king (ca. 910-28), 10
Trajanopolis, *6*
Transnistria, 44; *44*
Transylvania, 3, 5, 9, 15, 21, 29, 34, 35, 37, 38, 42, 45, 47, 48; *1, 8, 9, 11, 13, 14, 15, 18, 19, 21, 22, 24, 25, 26, 27, 28, 29, 31, 34, 35, 37, 38, 42, 44, 48*
"Transylvanian Question," 38; *38*
Travnik, *23, 32*
Trianon, treaty of (1920), 35, 37, 38, 42
Trieste, *35, 36, 43, 44, 45, 47, 48*
Trikala, *27, 40, 43, 46*
Tripolis, *40, 42, 43, 45, 46, 48*
Trogir, *10*
Trpimirović, Zvonimir, Croatian king (1075-90), 10
Truman, Harry S., American president (1945-53), 46
"Truman Doctrine," 46
Trumbić, Ante, 34, 36
Tsaribrod, *39*
Tudjman, Franjo, Croatian president (1991-99), 49
Tundzha River, *39*
Turkey, 4, 5, 35, 39, 40, 44, 48; *2, 3, 4, 5, 35, 39, 40, 42, 44, 46, 47, 48*
Turks, 2, 4, 5, 7, 8, 11, 14, 18, 19, 37, 38, 40, 48; *4;* Ottoman,

4, 5, 14, 17, 18, 19, 31, 33; Seljuks, 11, 12, 18
Tŭrnovo, 13, 19, 22, 29; *11, 12, 13, 14, 17, 18, 19, 22, 24, 26, 28, 29, 31, 34, 39, 45, 48*
Tŭrnovo, patriarchate of, 13, 18, 22
Turnu-Severin, *36, 37, 38*. *Also see* Severin
Tuzla, *32, 49*
Tvrtko I, Bosnian king-*ban* (1353-91), 16
Tzimiskes, John I, Byzantine emperor (969-76), 9
Uglješa, Jovan, 18, 19
Ukraine, 4, 7, 21; *2, 3, 4*
Ukrainian Steppe, 15, 21; *6*
Ukrainians, 2, 4, 37
Ulcinj, *41*
Una River, *10, 16*
United Nations (UN), 49
United States (America), 35, 41, 44, 45, 46, 47, 49, 50
Urban, 20
Urban II, pope (1088-99), 11, 12
Ustaše, 36, 42, 43, 44
Utigurs, *6*
Varaždin, *23, 32, 49*
Vardar/Axios River, 1, 3; *1, 30, 36, 40, 46, 50*
Varna (Odessos), 19; *6, 7, 8, 11, 12, 14, 15, 18, 19, 22, 27, 28, 31, 33, 34, 35, 37, 39, 42, 43, 44, 45, 47, 48*
Veles, *50*
Venetians, 10, 12
Venice, 10, 12, 13, 19, 21; *2, 8, 9, 11, 12, 13, 14, 17, 18, 19, 21, 22, 24, 25, 26, 27, 28, 29, 31, 33, 34, 35, 42, 43, 44, 45, 47, 48*
Venizelos, Elevtherios, 34, 35, 40
Versailles (Paris), treaties of (1919-20), 35, 36, 38, 39, 41, 42
Via Ignatia, 6, 12
Victor Emmanuel III, Italian king (1900-46), 41, 42
Victoria, British queen (1837-1901), 27
Vidin, 18, 24; *7, 8, 11, 15, 17, 18, 19, 24, 25, 26, 27, 28, 36, 37, 39, 42, 45*
Vidovdan (St. Vitus's Day), 32
Vienna, 21, 23, 24; *2, 9, 11, 12, 13, 14, 18, 19, 21, 22, 24, 25, 26, 27, 28, 29, 31, 34, 35, 42, 43, 44, 45, 47, 48*
Vienna Award, Second (1940), 38, 42
Vinkovci, *23*
Virovitica, *23*
Višegrad, *23, 32*
Visoko, *16, 18*
Vijosë River, 1, 3; *1, 41*
Vladislav, Ivan, Bulgarian *tsar* (1015-18), 9
Vlahs, 2, 4, 5, 15; *9, 11*
Vlaicu, Vladislav I, Wallachian *voievod* (1364-77), 15
Vlorë (Valona), *6, 7, 17, 33, 34, 35, 40, 41, 42, 43, 44, 45, 48*
Vojvodina, 5, 23, 25, 35, 43, 44, 45, 49, 50; *1, 35, 44, 45, 47, 48, 49*
Vranje, *30, 50*
Vratsa, *39*
Vrbas River, *10, 16*
Vukan, Zetan prince, 17
Vukovar, 49; *32, 48, 49*
Wallachia, 3, 4, 7, 8, 9, 15, 19, 20, 21, 24, 25, 27, 37; *1, 8, 14, 15, 17, 18, 19, 21, 22, 24, 25, 35, 37, 38*
Warsaw Pact, 47, 48
West Germany, *45, 47*. *Also see* Germany
Wied, William of, Albanian prince (1914), 41
Wilson, Woodrow, American president (1913-21), 35
Wittelsbach, Otto (Othon) I, Greek king (1832-62), 25
World War I, 28, 31, 32, 34, 35, 37, 41, 44; *34*
World War II, 38, 39, 40, 42, 43, 44, 46, 49; *43, 44*
Yalta, conference of (1945), 45, 46
Yantra River, *39*
"Young Bosnia," 32
Young Turk Revolution (1908), 31
Young Turks, 31, 33
Ypsilantis, Alexander, 25
"Yugoslav Group," 34
Yugoslav London Committee, 34, 36
Yugoslav National Army (JNA), 49, 50
Yugoslavia, 2, 3, 5, 34, 35, 36, 39, 40, 41, 43, 44, 45, 46, 47, 48, 49, 50; *2, 3, 4, 5, 36, 37, 38, 39, 40, 41, 42, 43, 45, 46, 47, 48, 49, 50*. *Also see* Kingdom of Serbs, Croats, and Slovenes
"Yugoslavism," 32, 34, 36, 49
Zadar, 13; *8, 10, 11, 12, 13, 14, 16, 18, 19, 21, 22, 23, 32, 33, 35, 36, 42, 43, 44, 45, 47, 48, 49*
Zagreb, 2, 23, 36; *2, 9, 10, 11, 12, 13, 14, 16, 18, 19, 21, 22, 23, 24, 25, 26, 27, 28, 29, 31, 32, 33, 34, 35, 36, 42, 43, 44, 45, 47, 48, 49*
Zakhariadis, Nikos, 46
Zápolya, János, Transylvanian prince (1526-40), 21
Zealots, 18
Zemun, *23, 25, 26, 27, 28, 29, 36*
Zenica, *32*
Žepa, *49*
Zervas, Napoleon, 44
Zeta, 9, 11, 17, 18; *8, 9, 10, 16, 17*
Zhivkov, Todor, Bulgarian Communist leader (1954-89), 48
Zlatitsa, *39*
Zog (Zogolli/Zogu, Ahmed) I, Albanian king (1928-39), 41